FM 3-21.5
C1

Change 1
Field Manual
3-21.5

Headquarters
Department of the Army
Washington, DC, April 2006

Drill and Ceremonies

1. Change FM 3-21.5 (FM 22-5), 7 July 2003, as follows:

Remove old pages:	Insert new pages:
Contents v through vi	Contents v through vi
10-7 through 10-10	10-7 through 10-10
10-15 through 10-16	10-15 through 10-16
14-1 through 14-16	14-1 through 14-14
Glossary 1 through 2	Glossary 1 through 2
References 1	References 1
Index 1 through 4	Index 1 through 4

2. A star (*) marks new or changed material. A star (*) on the revised contents page could indicate a minor (single word) or entire paragraph change.

3. File this transmittal sheet in front of the publication.

DISTRIBUTION RESTRICTION: Approved for public release; distribution is unlimited.

By Order of the Secretary of the Army:

PETER J. SCHOOMAKER
General, United States Army
Chief of Staff

Official:

JOYCE E. MORROW
Administrative Assistant to the
Secretary of the Army

0608102

DISTRIBUTION:

Active Army, Army National Guard, and United States Army Reserve. To be distributed in accordance with the initial distribution number 110178, requirements for FM 3-21.5.

FIELD MANUAL
No. 3-21.5

HEADQUARTERS
DEPARTMENT OF THE ARMY
WASHINGTON, DC, 7 July 2003

DRILL AND CEREMONIES

CONTENTS

Page

DISTRIBUTION RESTRICTION: Approved for public release; distribution is unlimited.

*This publication supersedes FM 22-5, dated 8 December 1986.

Page

PREFACE

This field manual provides guidance for Armywide uniformity in the conduct of drill and ceremonies. It includes methods of instructing drill, teaching techniques, individual and unit drill, manual of arms for infantry weapons, and various other aspects of basic drill instruction.

This manual is designed for use by soldiers of all military occupational skills, to include the new soldier in the initial entry training environment. Since all situations or eventualities pertaining to drill and ceremonies cannot be foreseen, commanders may find it necessary to adjust the procedures to local conditions. However, with the view toward maintaining consistency throughout the Army, the procedures prescribed herein should be followed as closely as possible.

Personnel preparing to give drill instruction must be thoroughly familiar with Chapters 3 through 7 before attempting to teach material in Chapters 8 and 9.

For ease in distinguishing a preparatory command from a command of execution, the commands of execution appear in **BOLD CAP** letters and preparatory commands appear in ***Bold Italic*** letters. Reference to positions and movements appear in *Italics*.

AR 670-1 should be used as a reference for the proper wearing of uniforms and insignia.

The proponent of this publication is the U.S. Army Infantry School. Send comments and recommendations to doctrine@benning.army.mil or on DA Form 2028 directly to Commandant, U.S. Army Infantry School, ATTN: ATSH-ATD, Fort Benning, Georgia 31905-5410.

Unless this publication states otherwise, masculine nouns and pronouns do not refer exclusively to men.

This page intentionally left blank.

PART ONE
DRILL

"Gentlemen: you have now reached the last point. If anyone of you doesn't mean business let him say so now. An hour from now will be too late to back out. Once in, you've got to see it through. You've got to perform without flinching whatever duty is assigned you, regardless of the difficulty or the danger attending it. If it is garrison duty, you must attend to it. If it is meeting fever, you must be willing. If it is the closest kind of fighting, anxious for it. You must know how to ride, how to shoot, how to live in the open. Absolute obedience to every command is your first lesson. No matter what comes you mustn't squeal. Think it over - all of you. If any man wishes to withdraw he will be gladly excused, for others are ready to take his place."

Theodore Roosevelt, Remarks to Recruits, 1898

CHAPTER 1
INTRODUCTION

The purpose of drill is to enable a commander or noncommissioned officer to move his unit from one place to another in an orderly manner; to aid in disciplinary training by instilling habits of precision and response to the leader's orders; and to provide for the development of all soldiers in the practice of commanding troops.

1-1. HISTORY

Military history reveals that armies throughout the world participated in some form of drill. The primary value of drill, historically, is to prepare troops for battle. For the most part, the drill procedures practiced are identical to the tactical maneuvers employed on the battlefield. Drill enables commanders to quickly move their forces from one point to another, mass their forces into a battle formation that affords maximum firepower, and maneuver those forces as the situation develops.

a. In 1775, when this country was striving for independence and existence, the nation's leaders were confronted with the problem of not only establishing a government but also of organizing an army that was already engaged in war. From the "shot heard around the world," on 19 April 1775, until Valley Forge in 1778, Revolutionary forces were little more than a group of civilians fighting Indian-style against well-trained, highly disciplined British Redcoats. For three years, General George Washington's troops had endured many hardships—lack of funds, rations, clothing, and equipment. In addition, they had suffered loss after loss to the superior British forces. These hardships and losses mostly stemmed from the lack of a military atmosphere in country. Thus, an army was created with little or no organization, control, discipline, or teamwork.

b. Recognizing the crisis, General Washington, through Benjamin Franklin, the American Ambassador to France, enlisted the aid of a Prussian officer, Baron Friedrich von Steuben. Upon his arrival at Valley Forge on 23 February 1778, von Steuben, a former staff officer with Frederick the Great, met an army of several thousand half-starved, wretched men in rags. He commented that a European army could not be kept together in such a state. To correct the conditions that prevailed, he set to work immediately and wrote drill movements and regulations at night and taught them the following day to a model company of 120 men selected from the line.

c. Discipline became a part of military life for these selected individuals as they learned to respond to command without hesitation. This new discipline instilled in the individual a sense of alertness, urgency, and attention to detail. Confidence in himself and his weapon grew as each man perfected the fifteen 1-second movements required to load and fire his musket. As the Americans mastered the art of drill, they began to work as a team and to develop a sense of pride in themselves and in their unit.

d. Watching this model company drill, observers were amazed to see how quickly and orderly the troops could be massed and maneuvered into different battle formations. Officers observed that organization, chain of command, and control were improved as each man had a specific place and task within the formation. Later, the members of the model company were distributed throughout the Army to teach drill. Through drill, they improved the overall effectiveness and efficiency of the Army.

e. To ensure continuity and uniformity, von Steuben, by then a major general and the Army Inspector General, wrote the first Army field manual in 1779, The Regulations for the Order and Discipline of the Troops of the United States, commonly referred to as the Blue Book. The drill procedures initiated at Valley Forge were not changed for 85 years, until the American Civil War, and many of the drill terms and procedures are still in effect today.

f. Drill commands are about the same as at the time of the War of 1812, except that then the officers and noncommissioned officers began them by saying, "Take care to face to the right, right, face." Also, during the American revolutionary period, troops marched at a cadence of 76 steps a minute instead of the current cadence of 120 steps. Then units performed precise movement on the battlefield, and the army that could perform them best was often able to get behind the enemy, or on his flank, and thus beat him. Speed spoiled the winning exactness. Also, firearms did not shoot far or accurately in 1776, so troop formations could take more time to approach the enemy.

g. As armament and weaponry improved, drill had to adapt to new tactical concepts. Although the procedures taught in drill today are not normally employed on the battlefield, the objectives accomplished by drill—professionalism, teamwork, confidence, pride, alertness, attention to detail, esprit de corps, and discipline—are just as important to the modern Army as they were to the Continental Army.

1-2. MILITARY MUSIC

The earliest surviving pictorial, sculptured, and written records show musical or quasimusical instruments employed in connection with military activity for signaling during encampments, parades, and combat. Because the sounds were produced in the open air, the instruments tended to be brass and percussion types. Oriental, Egyptian, Greek, Roman, and American Indian chronicles and pictorial remains show trumpets and drums of many varieties allied to soldiers and battles.

a. **Bugle Calls.** Bugle calls are used in U.S. military service as the result of the Contintental Army's contact with the soldiers and armies from Europe during the revolutionary period. After the American Revolution, many of the French (and English) bugle calls and drum beats were adopted by the United States Army.

b. **Attention.** This is taken from the British "Alarm," at which call the troops turned out under arms.

c. **Adjutant's Call.** The adjutant's call indicates that the adjutant is about to form the guard, battalion, or regiment.

d. **To the Color.** The old cavalry call, "To the Standard," in use from about 1835, was replaced by the present call of "To the Color."

e. **National Anthem.** "The Star Spangled Banner" officially became the National Anthem by law on 3 March 1931, in Title 36, United States Code 170.

f. **Sound Off.** The band, in place, plays "Sound Off" (three chords). It then moves forward and, changing direction while playing a stirring march, troops the line and marches past the soldiers in formation, then returns to its post. Upon halting, the band again plays three chords.

g. **Retreat.** Retreat is the ceremony that pays honors to the national flag when it is lowered in the evening.

h. **Official Army Song.** The official Army song, "The Army Goes Rolling Along," was formally dedicated by the Secretary of the Army on Veterans Day, 11 November 1956, and officially announced on 12 December 1957 (AR 220-90). In addition to standing while the National Anthem is played, Army personnel stand at attention whenever the official song is played. Although there is no Department of the Army directive in this regard, commanders, other officers, and other personnel can encourage the tribute to the Army by standing at attention when the band plays "The Army Goes Rolling Along."

CHAPTER 2
DRILL INSTRUCTIONS

"Troops who march in an irregular and disorderly manner are always in great danger of being defeated."

Vegetius: De Re Militari: A.D. 378

Section I. INSTRUCTIONAL METHODS

The progress made by individuals and units in precise and unified action in drill is determined by the following:

- The methods of instruction and the thoroughness of the instructor.
- The organization of soldiers into units of the most effective instructional size.

The three methods of instruction used to teach drill to soldiers are: step-by-step, talk-through, and by-the-numbers. The character of the particular movement being taught will indicate the most effective method to use. As a rule, marching movements are taught using the step-by-step method. Movements that require numerous or simultaneous actions by an individual or unit are best taught using the talk-through method. Movements that have two or more counts are taught using the by-the-numbers method.

To ensure that a soldier develops satisfactory proficiency during the time allotted, and to ensure a complete and consistent presentation by the drill instructor, each movement (regardless of the method used) should be presented using three teaching stages: explanation, demonstration, and practice.

2-1. EXPLANATION

In the explanation stage, the instructor must:

- Give the name of the movement.
- Give the practical use for the movement.
- Give the command(s) for the movement and explain its elements: the preparatory command and the command of execution. He must also discuss the command(s) necessary to terminate the movement. (Supplementary commands are discussed where appropriate in the explanations.)

This is how an instructor teaches a marching movement using the first teaching stage.

"Platoon, **ATTENTION. AT EASE.** The next movement (position), which I will name, explain, and have demonstrated, and which you will conduct practical work on, is the *30-Inch Step From the Halt.* This movement is used to march an element from point A to point B in a uniform manner. The command to execute this movement is ***Forward,*** **MARCH.** This is a two-part command: ***Forward*** is the preparatory command, and **MARCH** is the command of execution. The command to terminate this movement is **HALT. HALT** is also a two-part command when preceded by a preparatory command such as ***Squad*** or ***Platoon***. I will use ***Demonstrator*** as the preparatory command and **HALT** as the command of execution. When given, these commands are as follows: ***Forward,*** **MARCH;** ***Demonstrator,*** **HALT."**

2-2. DEMONSTRATION

In the demonstration stage, the instructor may use the step-by-step, talk-through, or by-the-numbers methods of instruction.

NOTE: The instructor may demonstrate the movement himself, modifying his position when necessary to maintain eye-to-eye contact with personnel being instructed.

a. **Step-by-Step Method of Instruction.** In the step-by-step method of instruction, the explanation and demonstration are combined, and the movements are taught one step at a time.

NOTE: The letters **P, I, C,** or **A** have been added to the end of certain paragraphs to help the reader understand the five-step process used in all marching movements known as the **PICAA** effect. Put simply, the **P**reparatory command, the **C**ommand of execution and the **A**ction step—executing the movement—are all given or executed when the same foot strikes the marching surface. The **I**ntermediate step and **A**dditional step are executed with the other foot.

(1) The instructor explains that on the command of execution, the demonstrator takes only one step and then stops in position until the command ***Ready,*** **STEP** (for the next step) is given. While the demonstrator is stopped in position, the instructor makes on-the-spot corrections and explains the actions to be taken on the next step. The instructor then has the demonstrator execute the movement at normal cadence.

(2) This is how an instructor teaches the demonstration stage when using the step-by-step method of instruction:

- *"**Demonstrator, POST.*** I will use the step-by-step method of instruction. On the preparatory command ***Forward*** of ***Forward,*** **MARCH,** without noticeable movement, shift the weight of the body onto the right foot. ***Forward.***"

- "On the command of execution **MARCH** of ***Forward,*** **MARCH,** step forward 30 inches with the left foot. The head, eyes, and body remain as in the *Position of Attention.* The arms swing in natural motion, without exaggeration and without bending the elbows, about 9 inches straight to the front and 6 inches straight to the rear of the trouser seams. The fingers and thumbs are curled as in the *Position of Attention,* just barely clearing the trousers. **MARCH.**"

- "On the command of execution **STEP** of ***Ready,*** **STEP,** execute a 30-inch step with the trail foot. Once again, ensure that the head, eyes, and body remain as in the *Position of Attention,* and that the arms swing naturally, without exaggeration and without bending the elbows, about 9 inches straight to the front and 6 inches straight to the rear of the trouser seams. The fingers and thumbs are curled, as in the *Position of Attention,* barely clearing the trousers. ***Ready,*** **STEP.** Notice that there are two steps explained: one from the *Halt* and one while marching. *"*

- "The command to terminate this movement is **HALT.** The preparatory command ***Demonstrator*** of ***Demonstrator,*** **HALT,** may be given as either foot strikes the marching surface. However, the command of execution **HALT** of ***Demonstrator,*** **HALT,** must be given the next time that same foot strikes the marching surface. The *Halt* is executed in two counts."

- "On the command of execution **STEP** of ***Ready,*** **STEP,** execute a 30-inch step with the trail foot. When that foot strikes the marching surface, the demonstrator will receive the **preparatory** command ***Demonstrator*** of ***Demonstrator,*** **HALT.** ***Ready, STEP. Demonstrator.*** *"* **(P—step 1 of the PICAA process)**

- "On the command of execution **STEP** of ***Ready,*** **STEP,** execute a 30-inch step with the trail foot. This is the **intermediate** (or thinking) step required between the preparatory command and the command of execution. ***Ready,*** **STEP."** **(I—step 2 of the PICAA process)**

- "On the command of execution **STEP** of ***Ready,*** **STEP,** execute a 30-inch step with the trail foot. When the foot strikes the marching surface, the demonstrator will receive the **command of execution HALT** of ***Demonstrator,*** **HALT.** ***Ready,*** **STEP. HALT.** The *Halt* is executed in two counts." **(C—step 3 of the PICAA process)**

- "On the command of execution **STEP** of ***Ready,*** **STEP,** execute a 30-inch step with the trail foot, this being the **additional** step required after the command of execution. ***Ready,*** **STEP."** **(A—step 4 of the PICAA process)**

- "On the command of execution **STEP** of ***Ready,*** **STEP,** bring the trail foot alongside the lead foot, reassuming the *Position of Attention,* thus terminating this movement. ***(Ready,*** **STEP. RE-FORM.)** At normal cadence, this movement would look as follows: ***Forward,*** **MARCH.** ***Demonstrator,*** **HALT. AT EASE."** **(A—step 5 of the PICAA process)**

- "Platoon, what are your questions pertaining to this movement when executed at normal cadence or when using the step-by-step method of instruction? (Clarify all questions.)"

- "***Demonstrator,*** **ATTENTION.** You will now become my assistant instructor. **FALL OUT."**

NOTE: Notice that when marching, there are five steps in the step-by-step method of instruction: 1 - **P**reparatory command step; 2 - **I**ntermediate step; 3 - **C**ommand of execution step; 4 - **A**dditional step; and 5 – **A**ction step.

b. **Talk-Through Method of Instruction.** In this method the explanation and demonstration are combined. Each movement. or action by the individual is executed as it is orally described.

(1) The instructor simply tells the demonstrator how and what he wants him to do. The demonstrator executes the movement as the instructor describes it. Then the instructor has the demonstrator execute the movement at normal cadence.

(2) This is how an instructor teaches the demonstration stage when using the talk-through method of instruction:

- ***"Demonstrator,*** **POST.** I will use the talk-through method of instruction. On the command of execution **ATTENTION** or **FALL IN**, sharply bring the heels together and on line, with the toes forming a 45-degree angle. Rest the weight of the body equally on the heels and balls of both feet. The legs are straight without locking the knees; the body is erect; hips level; chest, lifted and arched; and shoulders, square and even. Keep the head erect and face straight to the front with the chin drawn in so that the alignment of the head and neck is vertical. The arms hang straight without stiffness. The fingers are curled so that the tips of the thumbs are alongside and touching the first joint of the forefingers. Keep the thumbs straight along the seams of the trousers with the first joint of the fingers touching the trouser legs. Remain silent and do not move unless otherwise directed."

- "**AT EASE**. This position, executed at normal cadence, is as follows: ***Demonstrator,*** **ATTENTION. AT EASE. FALL IN. AT EASE."**

- Platoon, what are your questions pertaining to the *Position of Attention* when executed at normal cadence or when executed using the talk-through method of instruction?" (Clarify all questions.)

- ***Demonstrator,*** **ATTENTION.** You will be my assistant instructor, **FALL OUT.**"

NOTE: When teaching squad, platoon, or company drills, this method should be modified so that individuals are talked into position rather than through the position.

c. **By-the-Numbers Method of Instruction.** ***By-The-Numbers*** is the command used to begin instructing one count at a time. ***Without-The-Numbers*** is the command used to terminate single-count instruction and return to normal cadence. The explanation and demonstration are combined. Movements are explained and demonstrated one count at a time.

(1) The instructor has the demonstrator execute the movement **by the numbers** (one count at a time). The instructor then has the demonstrator execute the movement at normal cadence.

(2) This is how an instructor teaches the demonstration stage using the by-the-numbers method of instruction:

- ***"Demonstrator,*** **POST.** I will use the by-the-numbers method of instruction. *Port Arms* from *Order Arms* is a two-count movement. On the command of execution **ARMS** of ***Port,*** **ARMS,** this being count one, grasp the barrel of the rifle with the right hand and raise the rifle diagonally across the body, ensuring that the right elbow remains down without strain. Simultaneously, grasp the rifle with the left hand at the handguard just forward of the slip ring, keeping the rifle about 4 inches from the belt. ***By-the-numbers, Port,*** **ARMS."**

- "On count two, release the grasp of the rifle with the right hand and regrasp the rifle at the small of the stock. Keep the rifle held diagonally across the body, about 4 inches from the belt, elbows drawn in at the sides, and ensure the right forearm is horizontal, thus assuming the position of *Port Arms.* ***Ready, TWO."***

- *"Order Arms* from *Port Arms is* a three-count movement. On the command of execution **ARMS** of ***Order,*** **ARMS,** this being count one, release the grasp of the rifle with the right hand and move the right hand up and across the body, approaching the front sight assembly from the right front, and firmly grasp the barrel without moving the rifle. Ensure the right elbow remains down without strain. ***Order,*** **ARMS."**

- "On count two, release the grasp of the rifle with the left hand, and with the right hand lower the rifle to the right side until it is about 1 inch off the marching surface. At the same time, guide the rifle into place with the left hand at the flash suppressor, fingers and thumb extended and joined, palm facing the rear. ***Ready,*** **TWO**."

- "On count three, sharply move the left hand to the left side as in the *Position of Attention* and, at the same time, gently lower the rifle to the marching surface with the right hand, assuming the position of *Order Arms.* ***Ready,*** **THREE."**

- "At normal cadence, these commands (movements) are as follows: ***Without-the-numbers, Port,*** **ARMS.** ***Order,*** **ARMS. AT EASE"**

- "Platoon, what are your questions pertaining to *Port Arms* from *Order Arms* and *Order Arms* from *Port Arms* when executed at normal cadence or when executed by-the-numbers?" (Clarify all questions.)

- ***"Demonstrator,*** **ATTENTION.** You will be my assistant instructor. **FALL OUT.**" (Notice that there is no ***Ready,*** **ONE** command.)

2-3. PRACTICE

The practice stage is executed in the same manner as the demonstration stage except that the instructor uses the proper designator for the size of his element. However, the instructor does not have his element execute the movement at normal cadence until his element has shown a satisfactory degree of proficiency executing the movement using the selected method of instruction.

Section II. INSTRUCTIONAL TECHNIQUES

Basic drill instruction includes line and U-formations, stationary and moving cadence counts, and dedicated instruction to units and individuals.

2-4. FORMATIONS

Instruction and practical work, in all phases of drill, can best be presented by using the U-formation or line formation. As a rule, stationary movements are taught using the U-formation; marching movements, to include squad, platoon, and company drills, are best taught from the line formation. Soldiers should be taught in the smallest formation possible to ensure individual attention; however, squad drill, whenever possible, should be taught to squad-size units, platoon drill to platoon-size units, and company drill to company-size units.

a. The U-formation is formed by the instructor commanding ***Count,*** **OFF.**

(1) On the command of execution "**OFF**" of ***Count,*** **OFF**, all personnel, except the right flank personnel, turn their head and eyes to the right and the right flank personnel count off with "**ONE.**" After the right flank soldiers have counted their number, the soldiers to their left count off with the next higher number and simultaneously turn their head and eyes to the front. All other members of the formation count off in the same manner until the entire formation has counted off. The instructor then commands ***First and second squads,*** **FALL OUT**.

(2) On the command of execution "**FALL OUT,**" the first squad faces to the right, and the second squad faces to the left. After the squads have faced, the instructor commands ***U-formation,*** **FALL IN.**

(3) On the command of execution **"FALL IN,"** the following actions happen simultaneously:

- The first squad executes a *Column Left* and takes one more step than the highest number counted when the formation counted off; halts, and faces to the left without command.
- The second squad marches forward until the lead man clears the last vacated position of the first squad and then executes a *Column Right*, taking three more steps than the highest number counted when the formation counted off; halts, and faces to the right without command.
- The third squad stands fast, and the fourth squad executes one *15-Inch Step to the Left.* The instructor ensures that the first and second squads are one step in

front of and one step outside the flanks of the third squad before commanding **AT EASE** beginning the instruction.

(4) To re-form the platoon in a line formation, the commands are: **FALL OUT (pause); FALL IN.** The members of the platoon execute in the reversre manner as prescibed above, taking the same number of steps.

b. When instructing using the line formation at normal interval, it is recommended that the first rank kneels (right knee), second rank executes one *15-Inch Step to the Left* and kneels (left knee), third rank stands fast, and the fourth rank takes one *15-Inch Step to the Left.* The instructor adjusts any additional ranks as necessary to ensure they are uncovered. This formation can quickly be formed by commanding ***Instructional Formation,*** **MARCH.** To re-form the unit into a line formation, the command is **FALL IN.** The members of the platoon execute in the reverse manner as prescribed above.

c. The instructor may find using the extended rectangular formation more suitable for use.

(1) To form the extended rectangular formation, the instructor commands ***Extend to the left,*** **MARCH**. All right flank soldiers stand fast and extend their arms at shoulder level, elbows locked, fingers and thumbs extended and joined, palms facing down. All other soldiers turn to the left and double-time forward. After taking a sufficient number of steps, the soldiers stop and face to the front and extend their arms in the same manner as the right flank soldiers, ensuring that there is about 12 inches between all soldiers. Dress is to the right and cover is to the front. The remainder of the body is in the position of attention.

(2) The instructor then commands ***Arms downward,*** **MOVE**. The soldiers lower their arms sharply to the sides as in the position of attention.

(3) The instructor then commands ***Left,*** **Face**. All soldiers execute a left face.

(4) The instructor then commands ***Extend to the left,*** **MARCH**. All soldiers execute as previosly described.

(5) The instructor then commands ***Arms downward,*** **MOVE**. All soldiers execute as previously described.

(6) The instructor then commands ***Right,*** **FACE**. All soldiers execute a right face.

(7) The instructor then commands ***From front to rear,*** **COUNT OFF**. (**COUNT OFF** is the entire command of execution). Each member of the first rank turns their head and eyes to the right and counts off with "**ONE,**" then faces back to the front. The remaining ranks execute in the same manner as the first rank counting off in the same manner as the first rank until the entire formation has counted off. The members of the last (rear) rank do not turn their head and eyes.

(8) The instructor then commands ***Even numbers to the left,*** **UNCOVER**. All even numbered soldiers jump squarely in the center of the interval, resuming the position of attention. The formation is now prepared for instruction.

(9) To return the formation to the original configuration, the instructor commands ***Assemble to the right,*** **MARCH**. All soldiers double-time to their original position in formation.

d. The instructor may find the circular formation more suitable for training.

(1) The instructor positions himself in front of the lead soldier in the lead squad and commands ***Circle formation,*** **FOLLOW ME**. The instructor double-times in a circle large enough for the formation and moves to the center of the circle. The members of the

lead squad follow at an arms length plus 6 inches (approximately 40 inches total). The squad leaders of the following squads begin double-timing at the correct distance from the last soldier in the preceding squad without command. After the entire formation is in a circle, the instructor comes to the position of attention and commands ***Quick time,*** **MARCH**. The soldiers begin marching normally. The instructor then gives directives to individual soldiers and corrects the distance between them until each member is at approximately double arm interval.

(2) The instructor then commands the formation to *Halt* using the appropriate preparatory command.

(3) The instructor then commands ***Left,*** **FACE**. All personnel face toward the instructor. The formation is now ready for instruction.

(4) To return the formation to its original configuration, the instructor commands the formation to attention and commands **FALL OUT**, and then **FALL IN**. On the command **FALL IN**, all soldiers return to their original position in formation.

NOTE: When conditions do not warrant this formation, soldiers may be directed to remain standing and to uncover. To assemble the unit, the command **FALL IN** is given.

2-5. INSTRUCTORS

When acting as instructors or assistant instructors, officers and noncommissioned officers go wherever they are needed. They correct mistakes and ensure steadiness and proper performance in ranks. When an individual shows that he is unable to execute the proper movements or assume the proper position, the instructor may physically assist the soldier.

2-6. CADENCE COUNTING

To enable soldiers to learn or maintain cadence and develop rhythm, the instructor should have them count cadence while marching.

a. To count cadence while marching at quick time, the instructor gives the preparatory command, ***Count Cadence,*** as the left foot strikes the marching surface, and the command of execution, **COUNT,** the next time the left foot strikes the marching surface. The soldier begins to count the next time the left foot strikes the marching surface and counts as each foot strikes the marching surface—**ONE, TWO, THREE, FOUR; ONE, TWO, THREE, FOUR.** To count cadence while double-timing, the procedures are basically the same, except the soldier only counts each time the left foot strikes the marching surface. To maintain cadence when marching, soldiers will be allowed to sing, or a drummer's beat may provide cadence.

b. For stationary movements of two or more counts, the instructor commands ***In Cadence, Right,*** **FACE.** The soldier simultaneously executes the first count of the movement on the command of execution and sounds off, **ONE**; as he executes the second count he sounds off, **TWO**.

NOTE: To halt execution of movements in cadence, the instructor commands ***Without Cadence,*** and resumes normal drill methods. For example, ***Without Cadence,*** **MARCH** when marching at *Count Cadence* or ***Without Cadence, Left,*** **FACE** for stationary movements.

c. As soldiers begin to master the art of drill, instructors try to create a spirit of competition among individuals and between units. Although repetition is necessary when teaching drill, instructors use competitive drill exercises to ensure that drill does not become boring or monotonous.

d. Mass commands are used to develop confidence and promote enthusiasm. They are effective in developing a command voice when instructing a leadership course. Procedures for using mass commands are discussed in Chapter 3.

This page intentionally left blank.

CHAPTER 3

COMMANDS AND THE COMMAND VOICE

"The spirit of discipline, as distinct from its outward and visible guises, is the result of association with martial traditions and their living embodiment."

B. H. Liddell Hart, Thoughts on War, 1944

Section I. COMMANDS

A drill command is an oral order of a commander or leader. The precision with which a movement is executed is affected by the manner in which the command is given.

3-1. RULES

The following rules for giving commands apply to the commander when the unit drills as a separate unit and not as part of a larger formation.

a. When at the *Halt,* the commander faces the troops when giving commands. On commands that set the unit in motion (marching from one point to another), the commander moves simultaneously with the unit to maintain correct position within the formation. (See Chapter 4, paragraphs 4-5c and 4-14, for more information on facing in marching.)

b. When marching, the commander turns his head in the direction of the troops to give commands.

c. Exceptions to these rules occur during ceremonies, which are discussed in Chapter 10.

d. When elements drill as part of a larger unit, the rules for supplementary commands apply (paragraph 3-4).

e. The commander gives the command **AS YOU WERE** to revoke a preparatory command that he has given. The command **AS YOU WERE** must be given before the command of execution. The commander cannot cancel the command of execution with **AS YOU WERE**. If an improper command is not revoked, the personnel execute the movement in the best manner possible.

3-2. TWO-PART COMMANDS

Most drill commands have two parts: the preparatory command and the command of execution. Neither part is a command by itself, but the parts are termed commands to simplify instruction. The commands ***Ready, Port,*** **ARMS,** and ***Ready, Aim,*** **FIRE,** are considered to be two-part commands even though they contain two preparatory commands.

a. The preparatory command states the movement to be carried out and mentally prepares the soldier for its execution. In the command ***Forward,*** **MARCH,** the preparatory command is ***Forward.***

b. The command of execution tells when the movement is to be carried out. In ***Forward,*** **MARCH**, the command of execution is **MARCH**.

c. To change direction of a unit when marching, the preparatory command and command of execution for each movement are given so they begin and end on the foot in the direction of the turn: ***Right Flank,*** **MARCH** is given as the right foot strikes the marching surface, and ***Left Flank,*** **MARCH** as the left foot strikes the marching surface. The interval between the preparatory command and the command of execution is **always** one step or count. The preparatory command and command of execution are **always** given when the same foot strikes the marching surface.

3-3. COMBINED COMMANDS

In some commands, the preparatory command and the command of execution are combined; for example, **FALL IN, AT EASE,** and **REST.** These commands are given without inflection and at a uniformly high pitch and loudness comparable to that for a normal command of execution.

3-4. SUPPLEMENTARY COMMANDS

Supplementary commands are oral orders given by a subordinate leader that reinforce and complement a commander's order. They ensure proper understanding and execution of a movement. They extend to the lowest subordinate leader exercising control over an element affected by the command as a separate element within the same formation.

a. A supplementary command may be a preparatory command, a portion of a preparatory command, or a two-part command. It is normally given between the preparatory command and the command of execution. However, when a command requires an element of a unit to execute a movement different from other elements within the same formation, or the same movement at a different time, subordinate leaders give their supplementary commands at the time prescribed by the procedures covering that particular movement.

> **EXAMPLE:**
>
> The platoon is in column formation, and the platoon leader commands ***Column of Twos From the Left*** (pause), **MARCH.** The first and second squad leaders command ***Forward;*** the third and fourth squad leaders command **STAND FAST**. On the command of execution **MARCH**, the first and second squads march forward. At the appropriate time, the squad leader (third squad) nearest the moving element commands ***Column Half Left,*** **MARCH** (for both remaining squads). As the third and fourth squad leaders reach the line of march, they automatically execute a *Column Half Right* and obtain normal distance behind the first and second squads.

b. A subordinate leader gives all supplementary commands over his right shoulder except when his command is based on the actions of an element on his left or when the subelement is to execute a *Column Left (Half Left)* or *Left Flank.* Giving commands over the left shoulder occurs when changing configuration or a formation, such as forming a file or a column of fours and re-forming. (See Chapter 7, paragraph 7-14, for more information on forming a file and re-forming.)

EXAMPLE:

The platoon is in column formation and is going to form a file to the left. The left flank squad leader will give the supplementary command ***Column,*** **LEFT** over the left shoulder, since the movement will be to the left. The other squad leaders will give the supplementary command **STAND FAST** over the left shoulder as their movement will be to the left and is based on an element to the left.

NOTE: When in formation at present arms and the preparatory command ***Order*** of ***Order,*** **ARMS** is given, subordinate leaders terminate their salute before giving their supplementary command.

c. Supplementary commands are not given by a subordinate leader for the combined commands **FALL IN, AT EASE, REST** or for mass drill when his element forms as part of a massed formation. However, supplementary commands are given when forming a mass or when forming a column from a mass. (See Chapter 4, note following paragraph 4-3d.)

d. Except for commands while in mass formation, platoon leaders give supplementary commands following all preparatory commands of the commander. When the preparatory command is ***Company,*** the platoon leaders immediately come to *Attention* and command ***Platoon.*** The company commander allows for all supplementary commands before giving the command of execution.

e. When no direction is given, the response is understood to be ***Forward;*** when no rate of march is given, the response is *Quick Time.* Normally, when a direction or rate of march is included in the preparatory command, only the direction or rate of march is given as a supplementary command.

3-5. DIRECTIVES

Directives are oral orders given by the commander that direct or cause a subordinate leader to take action.

a. The commander gives directives rather than commands when it is more appropriate for subordinate elements to execute a movement or perform a task as independent elements of the same formation.

b. Directives are given in sentence form and are normally prefaced by the terms **HAVE YOUR UNITS** or **BRING YOUR UNITS**.

EXAMPLE:

HAVE YOUR UNITS OPEN RANKS AND STACK ARMS; BRING YOUR UNITS TO PRESENT ARMS.

c. **TAKE CHARGE OF YOUR UNITS** is the only directive on which a commander relinquishes his command and on which salutes are exchanged.

Section II. THE COMMAND VOICE

A correctly delivered command will be understood by everyone in the unit. Correct commands have a tone, cadence, and snap that demand willing, correct, and immediate response.

3-6. VOICE CONTROL

The loudness of a command is adjusted to the number of soldiers in the unit. Normally, the commander is to the front and center of the unit and speaks facing the unit so that his voice reaches everyone.

a. The voice must have carrying power, but excessive exertion is unnecessary and harmful. A typical result of trying too hard is the almost unconscious tightening of the neck muscles to force sound out. This produces strain, hoarseness, sore throat, and worst of all, indistinct and jumbled sounds instead of clear commands. Ease is achieved through good posture, proper breathing, correct adjustment of throat and mouth muscles, and confidence.

b. The best posture for giving commands is the position of *Attention.* Soldiers in formation notice the posture of their leader. If his posture is unmilitary (relaxed, slouched, stiff, or uneasy), the subordinates will imitate it.

c. The most important muscle used in breathing is the diaphragm—the large muscle that separates the chest cavity from the abdominal cavity. The diaphragm automatically controls normal breathing and is used to control the breath in giving commands.

d. The throat, mouth, and nose act as amplifiers and help to give fullness (resonance) and projection to the voice.

3-7. DISTINCTIVENESS

Distinctiveness depends on the correct use of the tongue, lips, and teeth, which form the separate sounds of a word and group the sounds into syllables. Distinct commands are effective; indistinct commands cause confusion. All commands can be pronounced correctly without loss of effect. Emphasize correct enunciation (distinctiveness). To enunciate clearly, make full use of the lips, tongue, and lower jaw.

To develop the ability to give clear, distinct commands, practice giving commands slowly and carefully, prolonging the syllables. Then, gradually increase the rate of delivery to develop proper cadence, still enunciating each syllable distinctly.

3-8. INFLECTION

Inflection is the rise and fall in pitch and the tone changes of the voice.

a. The preparatory command is the command that indicates movement. Pronounce each preparatory command with a rising inflection. The most desirable pitch, when beginning a preparatory command, is near the level of the natural speaking voice. A common fault with beginners is to start the preparatory command in a pitch so high that, after employing a rising inflection for the preparatory command, it is impossible to give the command of execution with clarity or without strain. A good rule to remember is to begin a command near the natural pitch of the voice (Figure 3-1).

b. The command of execution is the command that indicates when a movement is to be executed. Give it in a sharper tone and in a slightly higher pitch than the last syllable of the preparatory command. It must be given with plenty of snap. The best way to develop a command voice is to practice.

c. In combined commands, such as **FALL IN** and **FALL OUT**, the preparatory command and command of execution are combined. Give these commands without inflection and with the uniform high pitch and loudness of a normal command of execution.

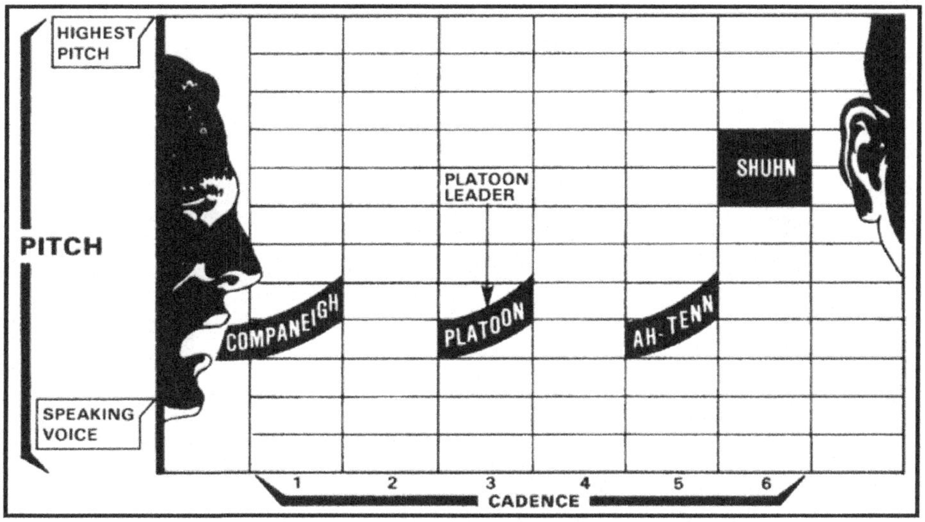

Figure 3-1. Diagram of a command.

3-9. CADENCE

Cadence, in commands, means a uniform and rhythmic flow of words. The interval between commands is uniform in length for any given troop unit. This is necessary so that everyone in the unit will be able to understand the preparatory command and will know when to expect the command of execution.

a. For the squad or platoon in *March,* except when supplementary commands need to be given, the interval of time is that which allows one step (or count) between the preparatory command and the command of execution. The same interval is used for commands given at the *Halt.* Longer commands, such as ***Right flank,*** **MARCH**, must be started so that the preparatory command will end on the proper foot, and leave a full count between the preparatory command and command of execution.

b. When supplementary commands are necessary, the commander should allow for one count between the preparatory command and the subordinate leader's supplementary command, and an additional count after the subordinate command but before the command of execution.

This page intentionally left blank.

CHAPTER 4
INDIVIDUAL DRILL

"Maintain discipline and caution above all things, and be on the alert to obey the word of command. It is both the noblest and the safest thing for a great army to be visibly animated by one spirit."

Archidamus of Sparta: To the Lacaedaenwnian expeditionary force departing for Athens, 431 B.C.

NOTE: The explanation of a movement that may be executed toward either flank is given in this chapter for only one flank. To execute the movement toward the opposite flank, substitute left for right or right for left in the explanation.

Section I. STATIONARY MOVEMENTS

This section contains most of the individual positions and stationary movements required in drill. These positions and the correct execution of the movement, in every detail, should be learned before proceeding to other drill movements.

Movements are initiated from the position of attention. However, some rest movements may be executed from other rest positions.

4-1. POSITION OF ATTENTION

Two commands can be used to put personnel at the *Position of Attention*:

- **FALL IN** is used to assemble a formation or return it to its original configuration.
- The two-part command for *Attention* is used for soldiers at a rest position.

Assume the *Position of Attention* on the command **FALL IN** or the command ***Squad (Platoon),*** **ATTENTION**.

a. To assume this position, bring the heels together sharply on line, with the toes pointing out equally, forming a 45-degree angle. Rest the weight of the body evenly on the heels and balls of both feet. Keep the legs straight without locking the knees. Hold the body erect with the hips level, chest lifted and arched, and the shoulders square.

b. Keep the head erect and face straight to the front with the chin drawn in so that alignment of the head and neck is vertical.

c. Let the arms hang straight without stiffness. Curl the fingers so that the tips of the thumbs are alongside and touching the first joint of the forefingers. Keep the thumbs straight along the seams of the trouser leg with the first joint of the fingers touching the trousers (Figure 4-1, page 4-2).

d. Remain silent and do not move unless otherwise directed.

NOTE: This position is assumed by enlisted soldiers when addressing officers, or when officers are addressing officers of superior rank.

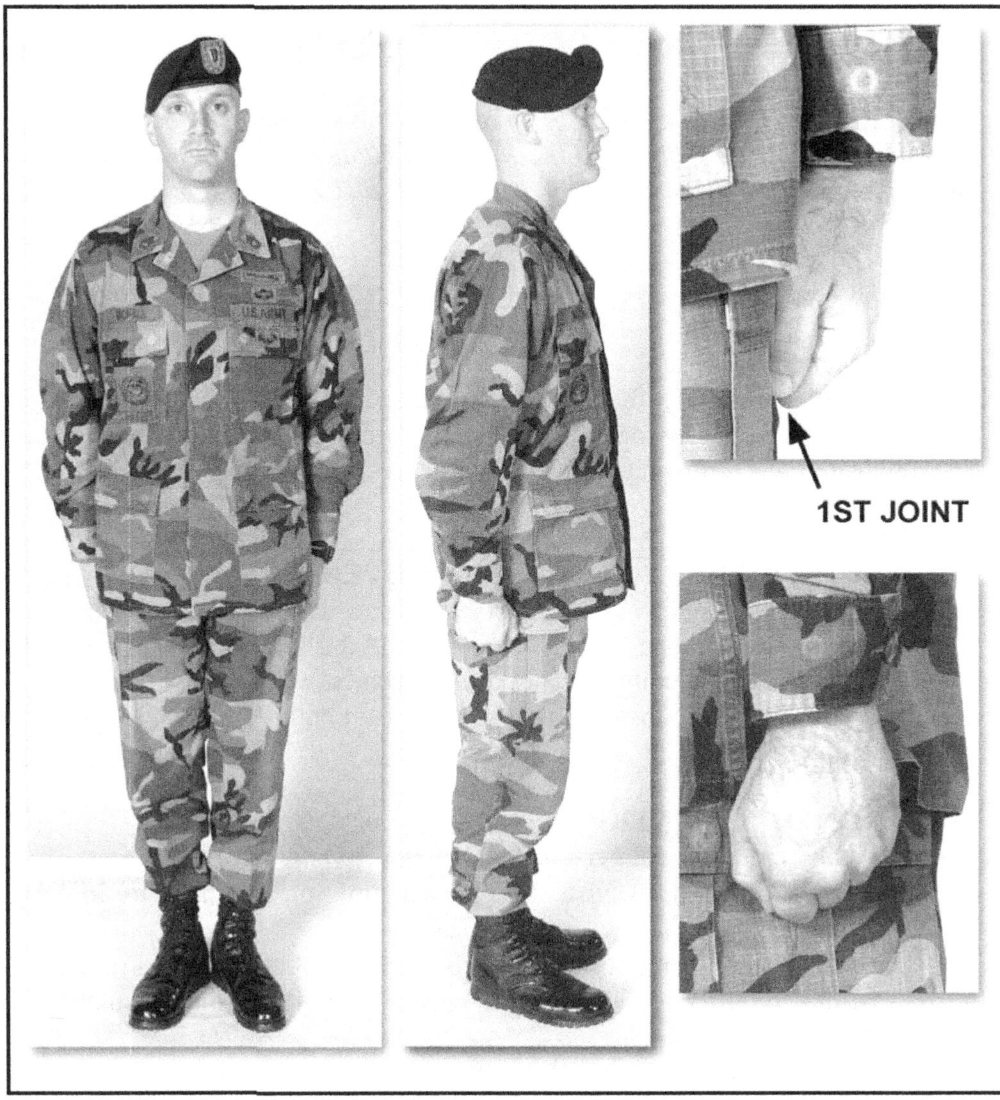

Figure 4-1. Position of Attention.

4-2. REST POSITIONS AT THE HALT

Any of the positions of rest may be commanded and executed from the *Position of Attention*.

a. **Parade Rest.** *Parade Rest is* commanded only from the *Position of Attention.* The command for this movement is ***Parade,*** **REST**.

(1) On the command of execution **REST**, move the left foot about 10 inches to the left of the right foot. Keep the legs straight without locking the knees, resting the weight of the body equally on the heels and balls of the feet.

(2) Simultaneously, place the hands at the small of the back and centered on the belt. Keep the fingers of both hands extended and joined, interlocking the thumbs so that the palm of the right hand is outward (Figure 4-2).

(3) Keep the head and eyes as in the *Position of Attention.* Remain silent and do not move unless otherwise directed. *Stand at Ease, At Ease,* and *Rest* may be executed from this position.

NOTE: Enlisted soldiers assume this position when addressing all noncommissioned officers or when noncommissioned officers address noncommissioned officers of superior rank.

Figure 4-2. Parade Rest.

b. **Stand At Ease.** The command for this movement is ***Stand at,*** **EASE**. On the command of execution **EASE**, execute *Parade Rest,* but turn the head and eyes directly toward the person in charge of the formation. *At Ease* or *Rest* may be executed from this position.

c. **At Ease.** The command for this movement is **AT EASE**. On the command **AT EASE**, the soldier may move; however, he must remain standing and silent with his right foot in place. The soldier may relax his arms with the thumbs interlaced. *Rest* may be executed from this position.

d. **Rest**. The command for this movement is **REST**. On the command **REST**, the soldier may move, talk, smoke, or drink unless otherwise directed. He must remain standing with his right foot in place. **AT EASE** must be executed from this position to allow soldiers to secure canteens, other equipment, and so forth.

NOTE: On the preparatory command for *Attention,* immediately assume *Parade Rest* when at the position of *Stand at Ease, At Ease,* or *Rest*. If, for some reason, a subordinate element is already at attention, the members of the element remain so and do not execute parade rest on the preparatory command, nor does the subordinate leader give a supplementary command.

4-3. FACING AT THE HALT

Five facing movements can be executed from the *Position of Attention:* ***Left (Right),*** **FACE,** ***Half Left (Half Right),*** **FACE,** and ***About,*** **FACE.**

NOTE: ***Half Left (Half Right),*** **FACE** should only be used in situations when a 90-degree facing movement would not face an element in the desired direction (for example, for a stationary element to face the direction of the flag to render honors [reveille or retreat]).

a. *Facing to the Flank* is a two-count movement. The command is ***Left (Right),*** **FACE.**

(1) On the command of execution **FACE,** slightly raise the right heel and left toe, and turn 90 degrees to the left on the left heel, assisted by a slight pressure on the ball of the right foot. Keep the left leg straight without stiffness and allow the right leg to bend naturally.

(2) On count two, place the right foot beside the left foot, resuming the *Position of Attention.* Arms remain at the sides, as in the *Position of Attention,* throughout this movement (Figure 4-3).

Figure 4-3. Left Face.

b. *Facing to the Rear* is a two-count movement. The command is ***About,*** **FACE.**

(1) On the command of execution **FACE,** move the toe of the right foot to a point touching the marching surface about half the length of the foot to the rear and slightly to the left of the left heel. Rest most of the weight of the body on the heel of the left foot and allow the right knee to bend naturally.

(2) On count two, turn to the right 180 degrees on the left heel and ball of the right foot, resuming the *Position of Attention.* Arms remain at the sides, as in the *Position of Attention,* throughout this movement (Figure 4-4, page 4-6).

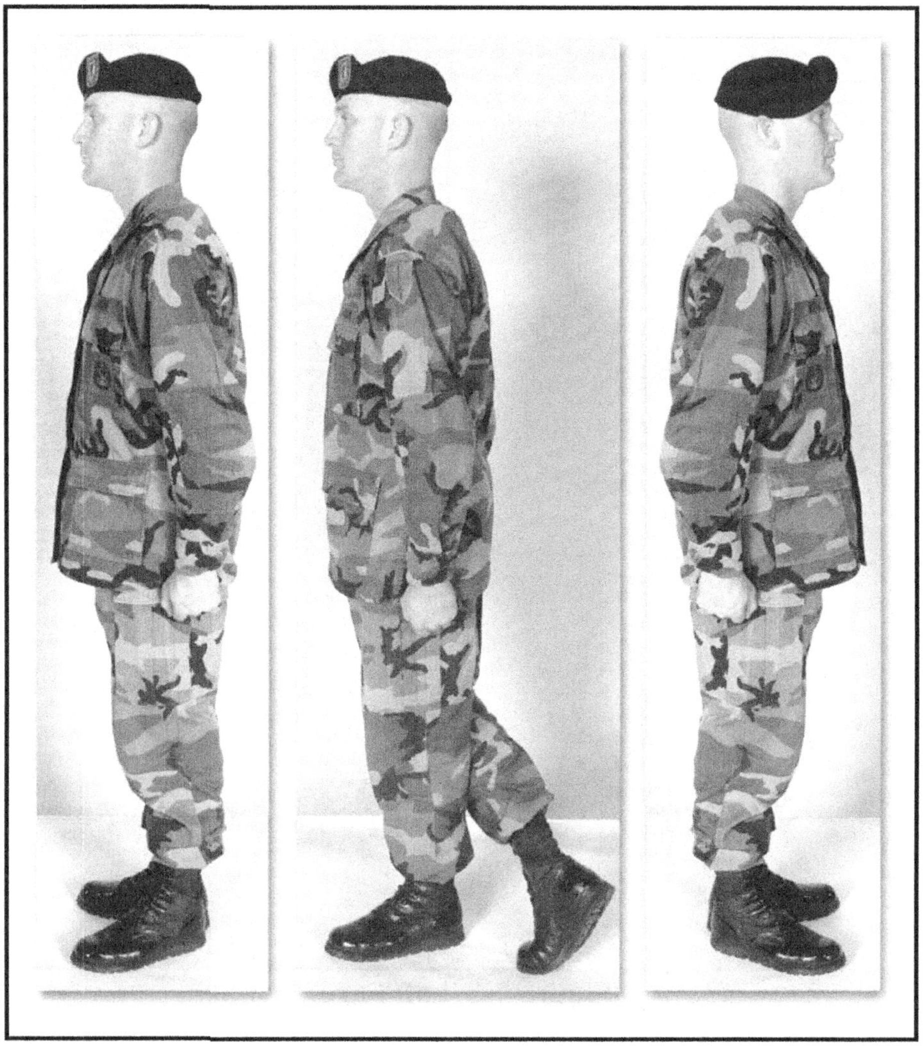

Figure 4-4. About Face.

NOTE: Throughout these movements, the remainder of the body remains as in the *Position of Attention.*

4-4. HAND SALUTE

The *Hand Salute* is a one-count movement. The command is ***Present,*** **ARMS.** The *Hand Salute* may be executed while marching. When marching, only the soldier in charge of the formation salutes and acknowledges salutes. When double-timing, an individual soldier must come to *Quick Time* before saluting.

a. When wearing headgear with a visor (with or without glasses), on the command of execution **ARMS,** raise the right hand sharply, fingers and thumb extended and joined, palm facing down, and place the tip of the right forefinger on the rim of the visor slightly to the right of the right eye. The outer edge of the hand is barely canted downward so that neither the back of the hand nor the palm is clearly visible from the front. The hand and wrist are straight, the elbow inclined slightly forward, and the upper arm horizontal (1, Figure 4-5).

b. When wearing headgear without a visor (or uncovered) and not wearing glasses, execute the *Hand Salute* in the same manner as previously described, except touch the tip of the right forefinger to the forehead near and slightly to the right of the right eyebrow (2, Figure 4-5).

c. When wearing headgear without a visor (or uncovered) and wearing glasses, execute the *Hand Salute* in the same manner as previously described, except touch the tip of the right forefinger to that point on the glasses where the temple piece of the frame meets the right edge of the right brow (3, Figure 4-5).

d. *Order Arms* from the *Hand Salute* is a one-count movement. The command is ***Order*, ARMS.** On the command of execution **ARMS,** return the hand sharply to the side, resuming the *Position of Attention.*

e. When reporting or rendering courtesy to an individual, turn the head and eyes toward the person addressed and simultaneously salute. In this situation, the actions are executed without command. The *Salute* is initiated by the subordinate at the appropriate time (six paces) and terminated upon acknowledgment. (See Appendix A for more information on saluting.)

Figure 4-5. Hand Salute.

Section II. STEPS AND MARCHING

This section contains all of the steps in marching of the individual soldier. These steps should be learned thoroughly before proceeding to unit drill.

4-5. BASIC MARCHING INFORMATION

This basic marching information pertains to all marching movements.

a. All marching movements executed from the *Halt* are initiated from the *Position of Attention.*

b. Except for *Route Step March* and *At Ease March,* all marching movements are executed while marching at *Attention.* Marching at *Attention* is the combination of the *Position of Attention* and the procedures for the prescribed step executed simultaneously.

c. When executed from the *Halt,* all steps except *Right Step* begin with the left foot. (See Chapter 3, paragraph 3-1a and paragraph 4-15a.)

d. For short-distance marching movements, the commander may designate the number of steps forward, backward, or sideward by giving the appropriate command: ***One step to the right (left),*** **MARCH;** or, ***Two steps backward (forward),*** **MARCH.** On the command of execution **MARCH,** step off with the appropriate foot, and halt automatically after completing the number of steps designated. Unless otherwise specified, when directed to execute steps forward, the steps will be 30-inch steps.

e. All marching movements are executed in the cadence of *Quick Time* (120 steps per minute), except the 30-inch step, which may be executed in the cadence of 180 steps per minute on the command ***Double Time,*** **MARCH.**

f. A step is the prescribed distance from one heel to the other heel of a marching soldier.

g. All 15-inch steps are executed for a short distance only.

4-6. THE 30-INCH STEP

To march with a 30-inch step from the *Halt,* the command is ***Forward,*** **MARCH.**

a. On the preparatory command ***Forward,*** shift the weight of the body to the right foot without noticeable movement. On the command of execution **MARCH,** step forward 30 inches with the left foot and continue marching with 30-inch steps, keeping the head and eyes fixed to the front. The arms swing in a natural motion, without exaggeration and without bending at the elbows, approximately 9 inches straight to the front and 6 inches straight to the rear of the trouser seams. Keep the fingers curled as in the *Position of Attention so* that the fingers just clear the trousers.

b. To *Halt* while marching, the command ***Squad (Platoon),*** **HALT** is given. The preparatory command ***Squad (Platoon)*** is given as either foot strikes the marching surface as long as the command of execution **HALT** is given the next time that **same foot** strikes the marching surface. The *Halt* is executed in two counts. After **HALT** is commanded, execute the additional step required after the command of execution and then bring the trail foot alongside the lead foot, assuming the *Position of Attention* and terminating the movement.

4-7. CHANGE STEP

This movement is executed automatically whenever a soldier finds himself out of step with all other members of the formation. It is only executed while marching forward with a 30-inch step. To change step, the command ***Change Step,*** **MARCH** is given as the right foot strikes the marching surface. On the command of execution **MARCH,** take one more step with the left foot, then in one count place the right toe near the heel of the left foot and step off again with the left foot. The arms swing naturally. This movement is executed automatically whenever a soldier finds himself out of step with all other members of the formation.

4-8. MARCHING TO THE REAR

This movement is used to change the direction of a marching element 180 degrees in a uniform manner. It is only executed while marching forward with a 30-inch step. To *March to the Rear*, the command ***Rear,*** **MARCH** is given as the right foot strikes the marching surface. On the command of execution **MARCH**, take one more step with the left foot, pivot 180 degrees to the right on the balls of both feet, and step off in the new direction taking a 30-inch step with the trail foot. Do not allow the arms to swing outward while turning.

4-9. REST MOVEMENT, 30-INCH STEP

Rest movements with the 30-inch step include *At Ease March* and *Route Step March.*

a. **At Ease March.** The command ***At Ease,*** **MARCH** is given as either foot strikes the marching surface. On the command of execution **MARCH,** the soldier is no longer required to retain cadence; however, silence and the approximate interval and distance are maintained. ***Quick Time,*** **MARCH** and ***Route Step,*** **MARCH** are the only commands that can be given while marching at ease.

b. **Route Step March.** *Route Step March* is executed exactly the same as *At Ease March* except that the soldier may drink from his canteen and talk.

NOTE: To change the direction of march while marching at *Route Step* or *At Ease March*, the commander informally directs the lead element to turn in the desired direction. Before precision movements may be executed, the unit must resume marching in cadence. The troops automatically resume marching at *Attention* on the command ***Quick Time,*** **MARCH**, as the commander reestablishes the cadence by counting for eight steps. If necessary, soldiers individually execute change step to get back in step with the unit.

4-10. THE 15-INCH STEP, FORWARD/HALF STEP

Use the following procedures to execute the 15-inch step, forward/half step.

a. To march with a 15-inch step from the *Halt,* the command is ***Half step,*** **MARCH.** On the preparatory command ***Half step,*** shift the weight of the body to the right foot without noticeable movement. On the command of execution **MARCH,** step forward 15 inches with the left foot and continue marching with 15-inch steps. The arms swing as in marching with a 30-inch step.

b. To alter the march to a 15-inch step while marching with a 30-inch step, the command is ***Half step,*** **MARCH.** This command may be given as either foot strikes the marching surface. On the command of execution **MARCH,** take one more 30-inch step and then begin marching with a 15-inch step. The arms swing as in marching with a 30-inch step.

c. To resume marching with a 30-inch step, the command ***Forward,*** **MARCH** is given as either foot strikes the marching surface. On the command of execution **MARCH,** take one more 15-inch step and then begin marching with a 30-inch step.

d. The *Halt* while marching at the *Half Step* is executed in two counts, the same as the *Halt* from the 30-inch step.

e. While marching at the *Half Step,* the only commands that may be given are: ***Mark Time,*** **MARCH;** ***Forward,*** **MARCH;** ***Normal Interval,*** **MARCH;** and **HALT.**

4-11. MARCHING IN PLACE

To march in place, use the following procedures.

a. To march in place, the command ***Mark Time,*** **MARCH** is given as either foot strikes the marching surface and only while marching with a 30-inch or 15-inch step forward. On the command of execution **MARCH**, take one more step, bring the trailing foot alongside the leading foot, and begin to march in place. Raise each foot (alternately) 2 inches off the marching surface; the arms swing naturally, as in marching with a 30-inch step forward.

NOTE: While marking time in formation, the soldier adjusts position to ensure proper alignment and cover. The proper distance between soldiers while marching is one arm's length plus 6 inches (approximately 40 inches).

b. To resume marching with a 30-inch step, the command ***Forward,*** **MARCH** is given as either foot strikes the marching surface. On the command of execution **MARCH**, take one more step in place and then step off with a 30-inch step.

c. The *Halt* from *Mark Time* is executed in two counts, basically the same as the *Halt* from the 30-inch step.

4-12. THE 15-INCH STEP, RIGHT/LEFT

To march with a *15-Inch Step Right (Left)*, use the following procedures:

a. To march with a *15-Inch Step Right (Left)*, the command is ***Right (Left) Step,*** **MARCH**. The command is given only while at the halt. On the preparatory command of ***Right (Left) Step,*** shift the weight of the body without noticeable movement onto the left (right) foot. On the command of execution **MARCH**, bend the right knee slightly and raise the right foot only high enough to allow freedom of movement. Place the right foot 15 inches to the right of the left foot, and then move the left foot (keeping the left leg straight) alongside the right foot as in the *Position of Attention.* Continue this movement, keeping the arms at the sides as in the *Position of Attention.*

b. To *Halt* when executing *Right* or *Left Step,* the command is ***Squad (Platoon),*** **HALT**. This movement is executed in two counts. The preparatory command is given when the heels are together; the command of execution **HALT** is given the next time the heels are together. On the command of execution **HALT**, take one more step with the lead foot and then place the trailing foot alongside the lead foot, resuming the *Position of Attention.*

4-13. THE 15-INCH STEP, BACKWARD

To march backward using the 15-inch step, use the following procedures:

a. To march with a *15-Inch Step Backward*, the command is ***Backward,*** **MARCH**. The command is given only while at the *Halt.* On the preparatory command ***Backward,*** shift the weight of the body without noticeable movement onto the right foot. On the command of execution **MARCH**, take a 15-inch step backward with the left foot and continue marching backward with 15-inch steps. The arms swing naturally.

b. The *Halt* from *Backward March* is executed in two counts, basically the same as the *Halt* from the 30-inch step.

4-14. THE 30-INCH STEP, DOUBLE TIME

To *Double-Time* using the 30-inch step, use the following procedures:

a. To march in the cadence of 180 steps per minute with a 30-inch step, the command is ***Double Time,*** **MARCH**. It may be commanded while at the *Halt* or while marching at *Quick Time* with a 30-inch step.

b. When at the *Halt* and the preparatory command ***Double Time*** is given, shift the weight of the body to the right foot without noticeable movement. On the command of execution **MARCH**, raise the forearms to a horizontal position, with the fingers and thumbs closed, knuckles out, and simultaneously step off with the left foot. Continue to march with 30-inch steps at the cadence of ***Double Time***. The arms swing naturally to the front and rear with the forearms kept horizontal. (When armed, soldiers will come to *Port Arms* on receiving the preparatory command of ***Double Time.***) Guides, when at *Sling Arms,* will *Double-Time* with their weapons at *Sling Arms* upon receiving the directive **GUIDE ON LINE.**

c. When marching with a 30-inch step in the cadence of *Quick Time,* the command ***Double Time,*** **MARCH** is given as either foot strikes the marching surface. On the command of execution **MARCH**, take one more 30-inch step at *Quick Time,* and step off with the trailing foot, double-timing as previously described.

d. To resume marching with a 30-inch step at *Quick Time,* the command ***Quick time,*** **MARCH** is given as either foot strikes the marching surface. On the command of execution **MARCH**, take two more 30-inch steps at *Double Time,* lower the arms to the sides, and resume marching with a 30-inch step at *Quick Time.*

NOTE: *Quick Time, Column Half Left (Right)*, and *Column Left (Right)* are the only movements that can be executed while double-timing. Armed troops must be at *Port Arms* before the command ***Double Time,*** **MARCH** is given.

4-15. FACING IN MARCHING

Facings in Marching from the *Halt* are important parts of the following movements: alignments, column movements, inspecting soldiers in ranks, and changing from *Normal Interval* to *Double Interval* or *Double Interval* to *Normal Interval.*

a. For instructional purposes only, the command ***Face to the Right (Left) in Marching,*** **MARCH** may be used to teach the individual to execute the movement properly. On the preparatory command ***Face to the Right (Left) in Marching,*** shift the weight of the body without noticeable movement onto the right foot. On the command of execution **MARCH**, pivot to the right (left) on the ball of the right foot (90 degrees) and step off in the indicated direction with the left foot. Execute the pivot and step in one count, and continue marching in the new direction. (See paragraph 4-5c and Chapter 3, paragraph 3-1a.)

b. *Facing* to the *Half-Right (Half-Left) in Marching* from the *Halt* is executed in the same manner as *Facing to the Right (Left) in Marching* from a *Halt,* except the facing movement is made at a 45-degree angle to the right (left).

c. The *Halt* from *Facing in Marching* is executed in two counts, the same as the *Halt* from the 30-inch step.

This page intentionally left blank.

CHAPTER 5
INDIVIDUAL DRILL WITH WEAPONS

Feed 'em up and give 'em hell. Teach 'em where they are. Make 'em so mad they'll eat steel rather than get dressing from you. Make 'em hard but don't break 'em.

Laurence Stallings: What Price Glory? 1926

NOTE: For individual movements with weapons other than the M16-series rifle, see the following appendixes:

- M4-series carbine—Appendix B.
- M14 rifle—Appendix C.
- M1903/M1917—Appendix D.
- Specialty weapons (M203, M249, shotgun, pistol)—Appendix E.
- Sword and saber—Appendix F.

Section I. MANUAL OF ARMS—M16-SERIES RIFLE

This section contains procedures for executing the manual of arms with the M16-series rifle (Figure 5-1) in conjunction with individual and unit drill movements.

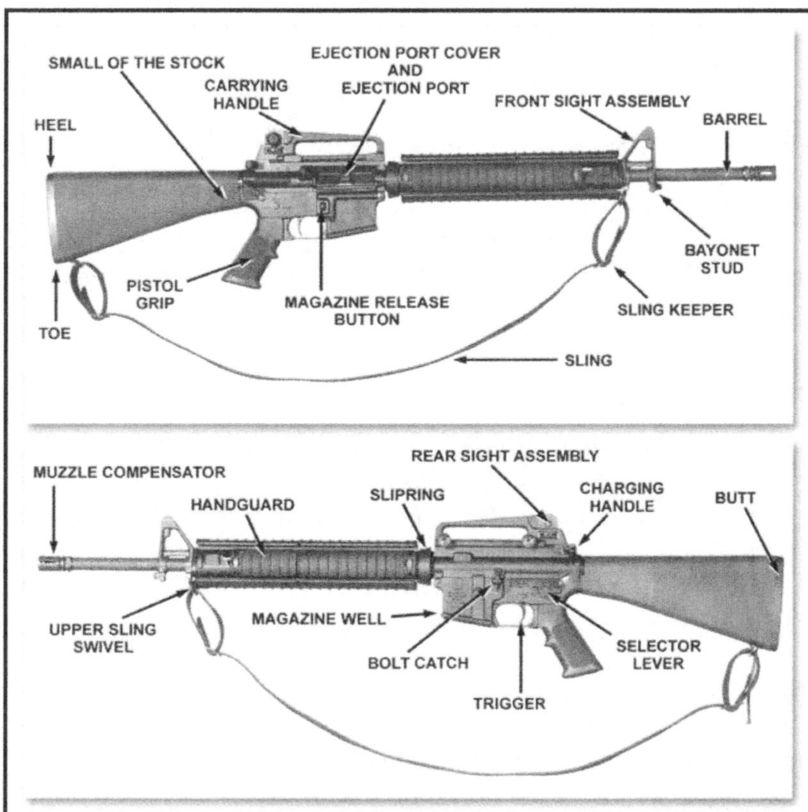

Figure 5-1. Nomenclature, M16-series rifle.

5-1. BASIC PROCEDURES

The following are basic procedures that apply to the M16-series rifle.

a. At the *Halt,* all movements are initiated from *Order Arms* or *Sling Arms,* which are the *Positions of Attention* with the rifle.

b. All precision movements are executed in quick-time cadence.

c. For drill purposes, the magazine is not carried in the rifle. When performing duty requiring the use of the magazine, the rifle is carried at *Sling Arms.*

d. The command ***Port,* ARMS** must be given before the command for *Double Time.*

e. *Facings,* alignments, and short-distance marching movements are executed from *Order Arms* or *Sling Arms.* When these movements are commanded while at *Order Arms,* automatically raise the rifle about 1 inch off the marching surface on the command of execution. When the movement has been completed, automatically return the rifle to *Order Arms.*

f. *Facing* movements are executed from *Order Arms* or *Sling Arms.* When a *Facing* is necessary to establish the direction of march, the *facing* movement is executed before the command for the manual of arms. After a marching movement has been completed, ***Order,* ARMS** or ***Sling,* ARMS** is commanded before the command for the *facing* movement.

g. ***Ready, Port,* ARMS** must be commanded following *Inspection Arms* and before any other movements can be commanded.

h. *Port Arms* is the key position assumed in most manual of arms movements from one position to another except *Right Shoulder Arms* from *Order Arms* and *Order Arms* from *Right Shoulder Arms.*

i. Manual of arms movements are a combination of the *Position of Attention* and the procedures for the prescribed movement. Most manual of arms movements are executed with the head, eyes, and body as in the *Position of Attention.*

j. *Sling Arms* is the appropriate position assumed to carry the rifle while marching in most situations. *Port Arms* is only used when slings are unavailable or unless specifically required by local conditions.

NOTES: 1. Paragraphs 5-2 through 5-10 refer to slings tight. However, all individual and unit drill movements may be executed with slings loose except when executing *Fix* and *Unfix Bayonets.*

2. The manual of arms movements for the M14, M1903/M1917, and M4 carbine are basically the same as for the M16. The movements for the M14 are fully described in Appendix B; the movements for the M1903/M1917 are fully described in Appendix C. The movements for the M4 carbine are fully described in Appendix D.

5-2. ORDER ARMS

To execute *Order Arms* use the following procedures:

a. Assume *Order Arms* on the command **FALL IN** or from *Parade Rest* on the command of execution **ATTENTION** (Figure 5-2).

b. At *Order Arms,* maintain the *Position of Attention* with the rifle. Place the butt of the rifle on the marching surface, centered on the right foot, with sights to the rear. The toe of the butt touches the foot so that the rear sight and pistol grip form a straight line to the front. Secure the rifle with the right hand in a "U" formed by the fingers (extended and joined) and thumb. Hold the rifle above the front sight assembly with the right thumb and forefinger pointed downward, and on line with the flat surface of the handguard. Keep the right hand and arm behind the rifle so that the thumb is straight along the seam of the trouser leg.

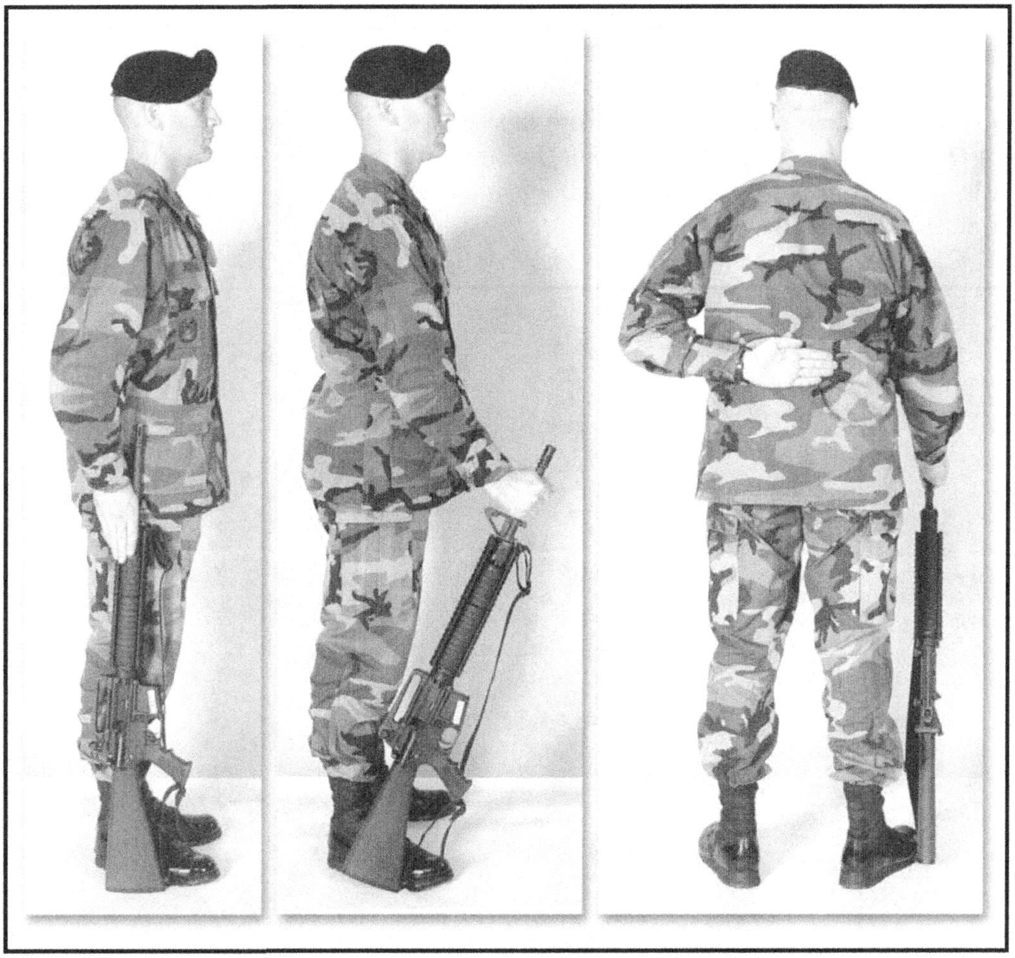

Figure 5-2. Order Arms and Parade Rest.

5-3. REST POSITION

The *Rifle Rest Positions* are commanded and executed the same as individual drill with the following additions:

a. On the command of execution **REST** of ***Parade,*** **REST**, thrust the muzzle forward, simultaneously changing the grip of the right hand to grasp the barrel, keeping the toe of the butt of the rifle on the marching surface and the right arm straight.

b. Execute *Stand at Ease* in the same manner as *Parade Rest* except turn the head and eyes toward the commander.

c. On the command **AT EASE** or **REST,** keep the butt of the rifle in place as in *Parade Rest.*

5-4. PORT ARMS

To execute *Port Arms* (Figure 5-3) use the following procedures:

a. *Port Arms* from *Order Arms* is a two-count movement. The command is ***Port,* ARMS.**

(1) On the command of execution **ARMS**, grasp the rifle barrel with the right hand and raise the rifle diagonally across the body, keeping the right elbow down (without strain). With the left hand, simultaneously grasp the handguard just forward of the slip ring so that the rifle is about 4 inches from the waist.

(2) On count two, regrasp the rifle at the small of the stock with the right hand. Hold the rifle diagonally across the body, about 4 inches from the waist, the right forearm horizontal, and the elbows close to the sides.

Figure 5-3. Port Arms from Order Arms.

b. *Order Arms* from *Port Arms* is a three-count movement (Figure 5-4). The command is **Order, ARMS.**

(1) On the command of execution **ARMS,** release the grasp of the right hand and move the right hand up and across the body to the right front of the front sight assembly, grasp the barrel firmly without moving the rifle, and keep the right elbow down without strain.

(2) On count two, move the left hand from the handguard and lower the rifle to the right side until it is about 1 inch from the marching surface. Guide the rifle to the side by placing the forefinger of the left hand at the flash suppressor, fingers and thumb extended and joined, palm to the rear.

(3) On count three, move the left hand sharply to the left side, lower the rifle gently to the marching surface, and resume the position of *Order Arms*.

Figure 5-4. Order Arms from Port Arms.

5-5. PRESENT ARMS

To execute *Present Arms* (Figure 5-5, page 5-6) use the following procedures:

a. *Present Arms* from *Order Arms* is a three-count movement. The command is ***Present,*** **ARMS.** On the command of execution **ARMS,** execute *Port Arms* in two counts. On count three, twist the rifle with the right hand so that the magazine well is to the front, and move the rifle to a vertical position with the carrying handle about 4 inches in front of and centered on the body. Lower the rifle until the left forearm is horizontal; keep the elbows in at the sides.

b. *Order Arms* from *Present Arms* is a four-count movement. The command is ***Order,* ARMS.** On the command of execution **ARMS,** return the rifle to *Port Arms.* Counts two, three, and four are the same as *Order Arms* from *Port Arms.*

c. *Port Arms* is assumed en route to or from *Present Arms* when going to or from *Right Shoulder* or *Left Shoulder Arms. Present Arms* from or to *Port Arms* is a one-count movement.

d. When rendering reports or courtesy to an individual from *Order Arms,* execute *Present Arms* and turn the head and eyes toward the individual addressed. *Order Arms* is executed automatically upon acknowledgment of the *Salute.*

e. When rendering courtesy to an individual with the rifle at *Right Shoulder, Left Shoulder,* or *Port Arms* and not in formation, execute *Present Arms.* Upon acknowledgment of the *Salute,* automatically return to the original position.

f. To render courtesy with the rifle at the *Carry* position, execute *Present Arms* from either the *Right-* or *Left-Hand Carry* position. Bring the rifle in the most convenient manner to *Port Arms,* and then execute *Present Arms* in one count. To assume the *Carry* position from *Present Arms,* execute *Port Arms* in one count and then return the rifle in the most convenient manner to the *Carry* position.

g. When double-timing (not in formation), come to *Quick Time* before rendering the courtesy.

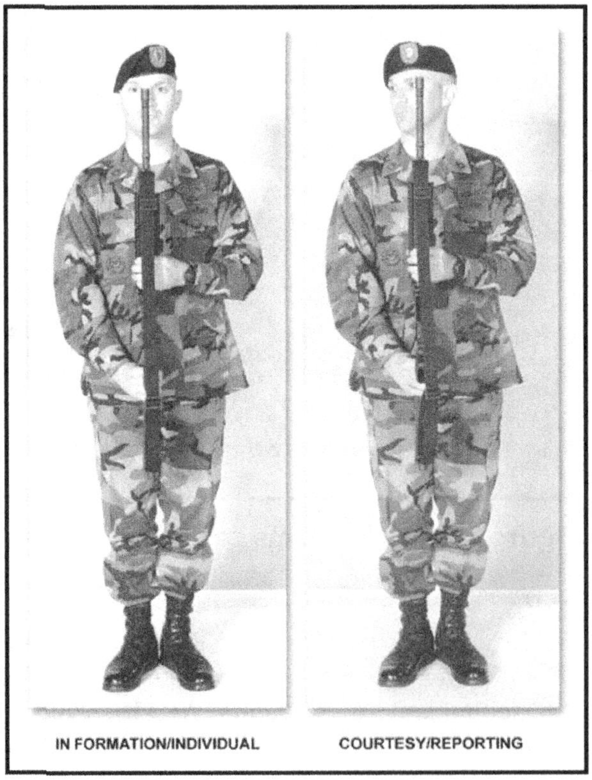

Figure 5-5. Present Arms.

5-6. INSPECTION ARMS

To execute *Inspection Arms* (Figure 5-6) use the following procedures:

a. *Inspection Arms* from *Order Arms* is a seven-count movement. The command is ***Inspection,* ARMS.**

(1) On the command of execution **ARMS,** execute *Port Arms* in two counts.

(2) On count three, move the left hand from the handguard and grasp the pistol grip, thumb over the lower portion of the bolt catch.

(3) On count four, release the grasp of the right hand, unlock the charging handle with the thumb, and sharply pull the charging handle to the rear with the thumb and forefinger. At the same time, apply pressure on the lower portion of the bolt catch, locking the bolt to the rear.

(4) On count five, without changing the grasp of the right hand, sharply push the charging handle forward until it is locked into position; then regrasp the rifle with the right hand at the small of the stock.

(5) On count six, remove the left hand, twist the rifle with the right hand so that the ejection port is skyward, regrasp the handguard with the left hand just forward of the slip ring, and visually inspect the receiver through the ejection port.

(6) On count seven, with the right hand, twist the rifle so that the sights are up and assume *Inspection Arms*.

Figure 5-6. Inspection Arms.

NOTE: The method for inspecting the rifle by an inspecting officer is explained in Chapter 8.

b. ***Ready, Port,*** **ARMS** is the only command given from *Inspection Arms.*

(1) On the command ***Ready,*** move the left hand and regrasp the rifle with the thumb and fingers forming a "U" at the magazine well and trigger guard, the thumb (without pressure) on the upper part of the bolt catch, fingertips placed below or under the ejection port cover.

(2) On the command ***Port,*** press the bolt catch and allow the bolt to go forward. With the fingertips, push upward and close the ejection port cover. Grasp the pistol grip with the left hand and place the left thumb on the trigger. On the command **ARMS,** pull the trigger and resume *Port Arms.*

5-7. RIGHT SHOULDER ARMS

To execute *Right Shoulder Arms* (Figure 5-7) use the following procedures:

a. *Right Shoulder Arms* from *Order Arms* is a four-count movement. The command is ***Right Shoulder,*** **ARMS.**

(1) On the command of execution **ARMS,** grasp the rifle barrel with the right hand and raise it diagonally across the body, keeping the right elbow down without strain. With the left hand, grasp the handguard just forward of the slip ring, ensuring that the weapon is about 4 inches from the waist.

(2) On count two, move the right hand from the barrel and grasp the heel of the butt between the first two fingers with the thumb and forefinger touching at the first joint.

(3) On count three (without moving the head), release the grasp of the left hand (without changing the grasp of the right hand), twist the rifle so that the sights are up, and place the weapon onto the right shoulder, moving the left hand to the small of the stock to guide the rifle to the shoulder. Keep the fingers and thumb (left hand) extended and joined with the palm turned toward the body. The first joint of the left forefinger touches the rear of the charging handle. Keep the left elbow down, and keep the right forearm horizontal with the right upper arm against the side and on line with the back.

(4) On count four, sharply move the left hand back to the left side as in the *Position of Attention.*

Figure 5-7. Right Shoulder Arms.

b. *Order Arms* from *Right Shoulder Arms* is a four-count movement. The command is ***Order,*** **ARMS.**

(1) On the command of execution **ARMS,** without moving the head and without changing the grasp of the right hand, press down quickly and firmly on the butt of the rifle with the right hand and twist the weapon (with the sights up), guiding it diagonally across the body and about 4 inches from the waist. Grasp the rifle with the left hand at the handguard just forward of the slip ring.

(2) On count two, move the right hand up and across the body, approaching from the right front of the front sight assembly, and firmly grasp the barrel without moving the rifle; keep the right elbow down without strain.

(3) Counts three and four are the same as from *Port Arms* to *Order Arms* (Figure 5-4, page 5-5).

5-8. LEFT SHOULDER ARMS

To execute *Left Shoulder Arms* (Figure 5-8) use the following procedures:

a. *Left Shoulder Arms* from *Order Arms* is a four-count movement. The command is ***Left shoulder,*** **ARMS.**

(1) On the command of execution **ARMS,** execute *Port Arms* in two counts.

(2) On count three, release the grasp of the left hand and (without moving the head) place the rifle on the left shoulder with the right hand (with the sights up), keeping the right elbow down. At the same time, regrasp the rifle with the left hand with the heel of the butt between the first two fingers and with the thumb and forefinger touching. The left forearm is horizontal, and the left upper arm is against the side and on line with the back.

(3) On count four, move the right hand to the right side as in the *Position of Attention.*

Figure 5-8. Left Shoulder Arms.

b. *Order Arms* from *Left Shoulder Arms* is a five-count movement. The command is ***Order,*** **ARMS.**

(1) On the command of execution **ARMS**, move the right hand up and across the body and grasp the small of the stock, keeping the right elbow down.

(2) On count two (without moving the head), release the grasp of the left hand and with the right hand move the rifle diagonally across the body (sights up) about 4 inches from the waist. At the same time, regrasp the handguard just forward of the slip ring with the left hand, and resume *Port Arms.*

(3) Counts three, four, and five are the same as *Order Arms* from *Port Arms* (Figure 5-4, page 5-5).

5-9. CHANGING POSITIONS

To change position with the M16-series rifle use the following procedures:

a. *Right Shoulder Arms* from *Port Arms* is a three-count movement. The command is ***Right Shoulder,*** **ARMS.** On the command of execution **ARMS,** release the grasp of the right hand and regrasp the rifle with the heel of the butt between the first two fingers, with the thumb and forefinger touching. Counts two and three are the same as counts three and four from *Order Arms.* When marching, the command is given as the right foot strikes the marching surface.

b. *Port Arms* from *Right Shoulder Arms* is a two-count movement. The command is ***Port,*** **ARMS.** On the command of execution **ARMS,** execute count one of *Order Arms* from *Right Shoulder Arms.* On count two, release the grasp of the right hand and regrasp the rifle at the small of the stock and come to *Port Arms.* When marching, the command is given as the right foot strikes the marching surface.

c. *Left Shoulder Arms* from *Port Arms* is a two-count movement. The command is ***Left Shoulder,*** **ARMS.** On the command of execution **ARMS,** execute *Left Shoulder Arms* in the same manner as counts three and four from *Order Arms.* When marching, the command is given as the left foot strikes the marching surface.

d. *Port Arms* from *Left Shoulder Arms* is a two-count movement. The command is ***Port,*** **ARMS.** On the command of execution **ARMS,** execute the first two counts of *Order Arms* from *Left Shoulder Arms.* When marching, the command is given as the left foot strikes the marching surface.

e. *Left Shoulder Arms* from *Right Shoulder Arms* is a four-count movement. The command is ***Left Shoulder,*** **ARMS.** On the command of execution **ARMS,** execute the first count the same as executing *Order Arms.* On count two, remove the right hand from the butt of the rifle and regrasp the small of the stock *(Port Arms).* Counts three and four are the same movements as from *Port Arms.* When marching, the command is given as the left foot strikes the marching surface.

f. *Right Shoulder Arms* from *Left Shoulder Arms* is a five-count movement. The command is ***Right Shoulder,*** **ARMS.** On the command of execution **ARMS,** execute *Port Arms* in two counts. Counts three, four, and five are the same as from *Port Arms.* When *marching,* the command is given as the right foot strikes the marching surface.

g. *Present Arms* from *Right Shoulder Arms* or *Left Shoulder Arms,* while in formation, is executed from the *Halt* only. The command is ***Present,*** **ARMS.** On the command of execution **ARMS,** come to *Port Arms* from either shoulder and then execute *Present Arms* (in one count) from *Port Arms.*

h. To resume *Right (Left) Shoulder Arms* from *Present Arms,* the command is ***Right (Left) Shoulder,*** **ARMS.** On the command of execution **ARMS,** execute *Port Arms* in one count and then execute the counts as prescribed from *Port Arms.*

NOTE: Experienced soldiers should be able to execute the 15-count manual of arms in unison from *Order,* to *Right Shoulder,* to *Left Shoulder,* to *Present,* to *Order Arms*. The command is ***Fifteen-Count Manual,*** **ARMS.**

5-10. FIX AND UNFIX BAYONETS

The command to *Fix* or *Unfix Bayonets* is given from *Order Arms* only. The movement is executed in a military manner but not in cadence.

NOTE: The bayonet scabbard is worn on the left side with the tip of the scabbard on line with the trouser leg seam and the barrel ring to the front.

a. To *Fix Bayonets,* the command is ***Fix,*** **BAYONETS**. On the command of execution **BAYONETS,** grasp the rifle barrel with the right hand, raise the rifle slightly, and place the butt of the rifle between the feet, with the magazine well to the front. Grasp the rifle barrel with the left hand and move the muzzle to the left front. With the right hand, unsnap the scabbard securing strap and withdraw the bayonet. Keeping the eyes on the bayonet point, turn the point skyward and attach the bayonet to the rifle. To engage the bayonet stud on the rifle with the base of the bayonet, grasp the handle, apply downward pressure until a click is heard, and then apply limited upward pressure to ensure that the bayonet is seated securely. Resnap the scabbard securing strap with the right hand and then come to *Order Arms.*

b. To *Unfix Bayonets,* the command is ***Unfix,*** **BAYONETS.** On the command of execution **BAYONETS,** grasp the rifle barrel with the right hand at the handle of the bayonet and place the rifle butt between the feet with the magazine well to the front. Move the muzzle to the left with the left hand and secure it. Unsnap the scabbard securing strap with the right hand, then grasp the bayonet handle with the left hand and release the bayonet from the rifle muzzle with the left hand. Keeping the eyes on the bayonet point, return the bayonet to the scabbard and insert it with the barrel ring facing to the front. Resnap the scabbard securing strap and come to *Order Arms.* For safety, if the bayonet is difficult to remove from the rifle, stick the bayonet point into the marching surface, bend over, and depress the catch mechanism with the left hand while pulling upward on the rifle with the right hand.

Section II. SLING ARMS—M16-SERIES RIFLE

This section contains the procedures for executing manual of arms movements while at *Sling Arms*.

5-11. BASIC PROCEDURES

These procedures apply to all movements while at *Sling Arms*.

a. Remain at *Sling Arms* during all rest movements.

b. All individual and unit drill movements can be executed at *Sling Arms* except *Double Time*, *Stack Arms* and *Fix* and *Unfix Bayonets*, which are executed from *Order Arms*.

c. When in formation at *Sling Arms*, execute the hand salute on the command ***Present,*** **ARMS.**

d. Platoon leaders and platoon sergeants carry their rifles at *Sling Arms* during all drills and ceremonies, and from this position they execute *Present Arms* only. They do not execute *Unsling Arms*. This also applies to squad leaders when squads drill as separate units.

e. When all members of a unit are carrying their rifles at *Sling Arms*, the platoon leaders and platoon sergeants execute *Present Arms* only. They do not execute *Unsling Arms*.

5-12. SLING ARMS

To execute *Sling Arms* use the following procedures:

a. From *Order Arms* with the sling(s) loose, the command for *Sling Arms* is ***Sling,*** **ARMS** (Figure 5-9, page 5-14). On the command of execution **ARMS,** grasp the rifle barrel with the right hand and raise it vertically. Grasp the sling near the upper sling swivel with the left hand, and release the right hand. Place the right hand and arm between the sling and rifle and place the sling over the right shoulder. Regrasp the sling with the right hand so that the wrist is straight, the right forearm is horizontal, the elbow is tight against the side, and the rifle is vertical. Release the grasp of the left hand and move it sharply to the left side as in the *Position of Attention*.

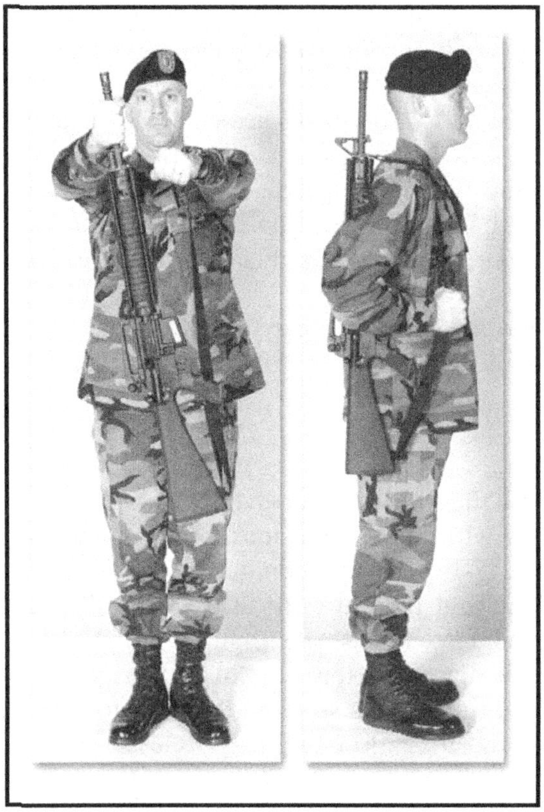

Figure 5-9. Sling Arms.

b. To return the rifle to *Order Arms* with the sling tight, the command is ***Adjust,*** **SLINGS** (Figure 5-10). On command of execution **SLINGS,** remove the rifle from the shoulder. Then, grasp the rifle barrel with the right hand and raise it vertically. With the left hand, place the butt of the rifle on the right hip and cradle it in the crook of the right arm. Use both hands to tighten the sling on the ejection port side. Grasp the rifle barrel with the right hand and guide the rifle to the *Order Arms* position (as previously described).

c. From *Order Arms* with sling(s) tight, the command is ***Sling,*** **ARMS.** On the command of execution **ARMS,** grasp the rifle barrel with the right hand and raise the rifle vertically. With the left hand, place the rifle butt on the right hip, cradle the rifle in the crook of the right arm, and use both hands to adjust the sling. Grasp the sling with the left hand near the upper sling swivel and execute *Sling Arms* (as previously described).

d. If an element is at *Order Arms* with the sling loose and the commander wants the sling to be tightened, he commands ***Adjust*** **SLINGS.** On the command of execution **SLINGS,** tighten the sling and guide the rifle to *Order Arms* (as previously described).

NOTE: Unless otherwise specified, armed elements of a formation fall-in at *Order Arms* with slings loose.

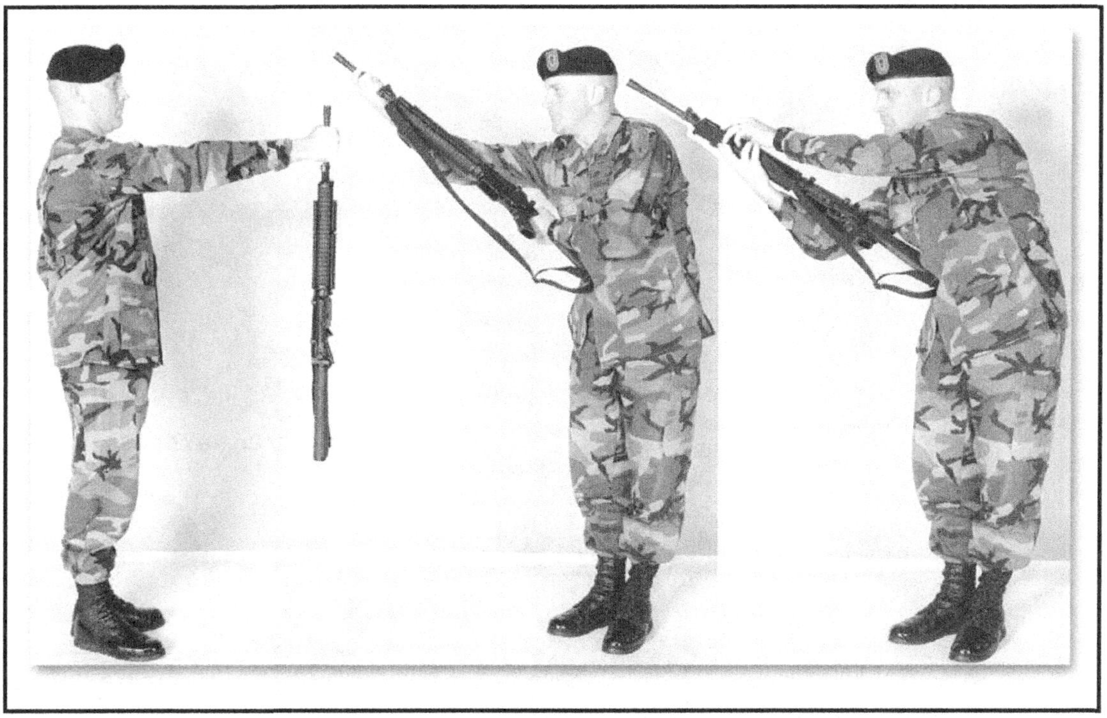

Figure 5-10. Adjust Slings.

e. To return the rifle to *Order Arms* with the sling loose, the command is ***Unsling,*** **ARMS.** On the command of execution **ARMS,** reach across the body with the left hand and grasp the sling at the right shoulder. Release the right hand grasp of the sling and remove the rifle from the shoulder. Grasp the rifle barrel with the right hand, release the left hand grasp of the sling, and guide the rifle to the *Order Arms* position (as previously described).

5-13. SALUTE AT SLING ARMS

To execute S*alute* while at *Sling Arms* (Figure 5-11, page 5-16) use the following procedures:

a. To *Salute* while at *Sling Arms,* the command is ***Present,*** **ARMS.** On the command of execution **ARMS,** reach across the body with the left hand and grasp the sling just above the right hand. Release the right hand and execute the *Hand Salute.*

b. To terminate the *Hand Salute,* the command is ***Order,*** **ARMS.** On the command of execution **ARMS,** lower the right hand sharply to the side as in the *Position of Attention* and then regrasp the sling at the original position. After grasping the sling with the right hand, release the left hand and return it sharply to the left side as in the *Position of Attention.*

c. When rendering reports or courtesy to an individual, the same rules apply for the *Hand Salute* (as explained in individual drill).

Figure 5-11. Salute while at Sling Arms.

NOTE: Individuals performing duty in congested areas, which would require frequent salutes, should carry the weapon at *Sling Arms*.

5-14. PORT ARMS FROM SLING ARMS

To execute *Port Arms* from *Sling Arms* (Figure 5-12) use the following procedures:

a. The command for this movement is ***Port,*** **ARMS.** On the command of execution **ARMS,** reach across the body with the left hand and grasp the sling at the shoulder. Lift the weapon (by the sling), swing it to the front of the body, and grasp the small of the stock with the right hand. Release the sling (left hand) and regrasp the weapon just forward of the slip ring. Ensure that the rifle is 4 inches from the belt and held diagonally across the body. Keep the elbows in at the sides and the right forearm horizontal.

b. To resume *Sling Arms,* the command is ***Sling,*** **ARMS.** On the command of execution **ARMS,** grasp the sling near the upper sling swivel with the left hand. Release the right hand and swing the rifle back onto the shoulder by placing the right arm between the sling and rifle, immediately resume the position of *Sling Arms.*

Figure 5-12. Port Arms from Sling Arms.

5-15. INSPECTION ARMS FROM SLING ARMS

To execute *Inspection Arms* from *Sling Arms* use the following procedures:

a. The command for this movement is ***Inspection,*** **ARMS.** On the command of execution **ARMS,** execute *Port Arms* and then execute counts three through seven in the same manner as for *Inspection Arms* from *Order Arms.*

b. ***Ready, Port,*** **ARMS** is executed in the same manner as explained in the manual of arms.

c. To resume *Sling Arms,* the procedures are the same as from *Port Arms.*

d. To execute *Inspection Arms* when the magazine is in the rifle, remove the magazine (just before count three) with the left hand and place it between the waist (left front) and the clothing. Return the magazine immediately after pulling the trigger and before resuming *Port Arms.* If the pistol belt is worn, the magazine will be placed between the pistol belt and the clothing (left front).

This page intentionally left blank.

CHAPTER 6
SQUAD DRILL

The [soldiers] must learn to keep their ranks, to obey words of command, and signals by drum and trumpet, and to observe good order, whether they halt, advance, retreat, are upon a march, or engaged with an enemy.

Niccolo Machiavelli: Arte della Guerra, 1520

Section I. FORMATIONS AND MOVEMENTS

This section describes the formations and movements of a squad. Individual drill movements and the manual of arms are executed as previously prescribed while performing as a squad member.

6-1. BASIC INFORMATION

The squad has two prescribed formations—line and column. However, the squad may be formed into a column of twos from a column formation. When the squad is in line, squad members are numbered from right to left; when in column, form front to rear. The squad normally marches in column, but for short distances it may march in line.

When the squad drills as a separate unit, the squad leader carries his weapon at *Sling Arms.* When the squad is in a line formation, the squad leader assumes a post three steps in front of and centered on the squad; when in a column or a column of twos, three steps to the left and centered on the squad. When the squad drills as part of a larger unit, the squad leader occupies the number one (base) position of the squad. He carries his weapon in the same manner as prescribed for other riflemen in the squad.

6-2. FORMING THE SQUAD

The squad normally forms in a line formation; however, it may re-form in column when each member can identify his exact position (equipment grounded) in the formation (Figure 6-1, page 6-2).

a. To form at normal interval, the squad leader comes to the *Position of Attention* and commands **FALL IN.** On the command **FALL IN,** the following actions occur simultaneously:

(1) Each member double-times to his position in the formation.

(2) The right flank man positions himself so that when the squad is formed it is three steps in front of and centered on the squad leader.

(3) The right flank man comes to the *Position of Attention* and raises his left arm laterally at shoulder level, elbow locked, fingers and thumb extended and joined, and palm facing down. He ensures that the left arm is in line with the body.

(4) The man to the immediate left of the right flank man comes to the *Position of Attention,* turns his head and eyes to the right, and raises his left arm in the same manner as the right flank man. He obtains proper alignment by taking short steps forward or backward until he is on line with the right flank man. He then obtains exact interval by taking short steps left or right until his shoulder touches the extended fingertips of the right flank man. As soon as the man to the left has obtained *Normal Interval,* each man

individually lowers his arm to his side, sharply turns his head and eyes to the front, and assumes the *Position of Attention.*

(5) The right flank man then sharply returns to the *Position of Attention.*

(6) All other members of the squad form in the same manner except that the left flank man does not raise his left arm.

NOTE: The right flank man raises his arm and looks straight to the front unless the squad is to align on an element to its right. If he is to align on an element to the right he turns his head and eyes to the right and aligns himself with that element.

b. To form at *Close Interval,* the formation is completed in the manner prescribed for *Normal Interval,* except that the command is ***At Close Interval,*** **FALL IN**. Squad members obtain *Close Interval* by placing the heel of the left hand on the left hip even with the waist, fingers and thumb joined and extended downward, and with the elbow in line with the body and touching the arm of the man to the left.

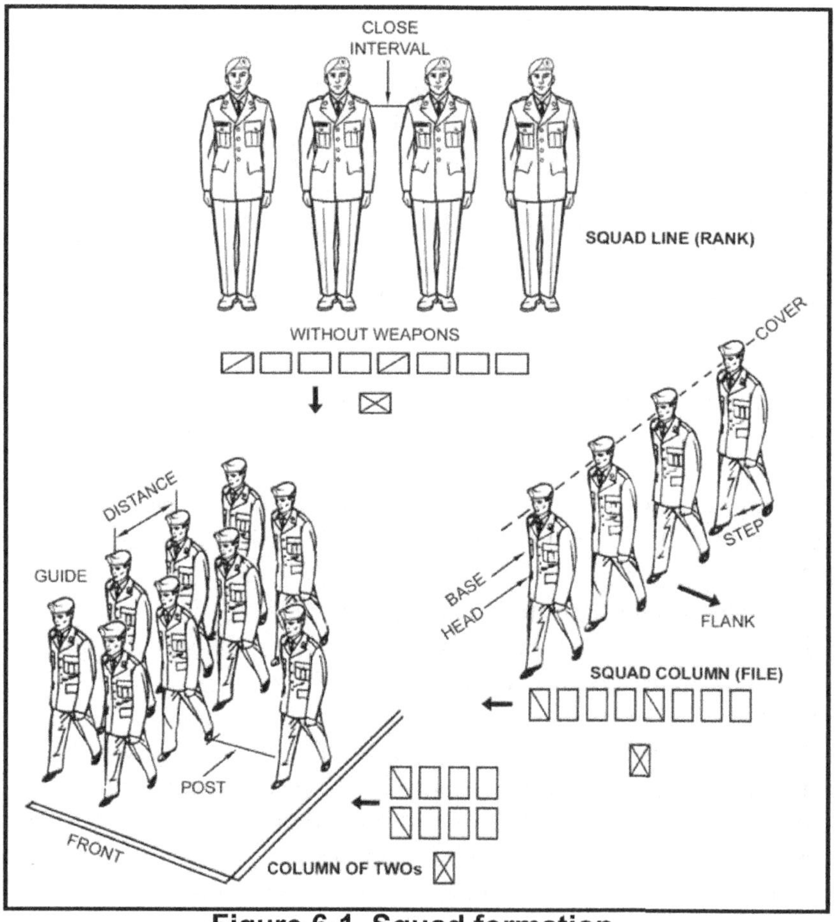

Figure 6-1. Squad formation.

c. To form in column, the squad leader faces the proposed flank of the column and commands ***In Column,*** **FALL IN**. On the command of execution **FALL IN**, squad members double-time to their original positions (grounded equipment) in formation and cover on the man to their front.

NOTE: The correct distance between soldiers in column formation is approximately 36 inches. This distance is one arm's length plus 6 inches

d. When armed, squad members fall in at *Order Arms* or *Sling Arms.* For safety, the commands ***Inspection,*** **ARMS;** ***Ready, Port,*** **ARMS;** and ***Order (Sling),*** **ARMS** are commanded at the initial formation of the day or when the last command is **DISMISSED** (Figure 6-1).

6-3. COUNTING OFF

The squad may count off in line or column formation. The command is ***Count,*** **OFF**.

a. When the squad is in a line formation, the counting is executed from right to left. On the command of execution **OFF**, each member, except the right flank man, turns his head and eyes to the right, and the right flank man counts off **"ONE."** After the man on the right counts off his number, the man to his left counts off with the next higher number and simultaneously turns his head and eyes to the front. All the other members execute count off in the same manner until the entire squad has counted off.

b. When the squad is in column formation, the counting is executed from front to rear. On the command of execution **OFF,** the soldier at the head of the column turns his head and eyes to the right and counts over his shoulder, **"ONE."** After counting off his number, he immediately comes to the *Position of Attention.* All other members count their numbers in sequence in the same manner as the number one man; the last man in the file does not turn his head and eyes to the right.

6-4. CHANGING INTERVAL WHILE IN LINE

To change interval while in line, use the following procedures:

NOTE: To ensure that each member understands the number of steps to take, the squad leader should command ***Count,*** **OFF** before giving any commands that cause the squad to change interval. Members do not raise their arms when changing interval.

a. To obtain *Close Interval* from *Normal Interval,* the command is ***Close Interval,*** **MARCH.** On the command of execution **MARCH,** the right flank man (number one man) stands fast. All men to the left of the number one man execute *Right Step March,* take one step less than their number (for example, number five man takes four steps), and *Halt.*

NOTE: The squad leader takes the correct number of steps to maintain his position of three steps in front of and centered on the squad.

b. To obtain *Normal Interval* from *Close Interval,* the command is ***Normal Interval,*** **MARCH.** On the command of execution **MARCH,** the right flank man stands fast. All men to the left of number one man execute *Left Step March,* take one step less than their number (for example, number nine man takes eight steps), and *Halt.*

c. To obtain *Double Interval* from *Normal Interval,* the command is ***Double Interval,*** **MARCH.** On the command of execution **MARCH,** the right flank man stands fast. All men to the left of number one man face to the left as in marching, take one 30-inch step less than their number (for example, number seven man takes six steps), *Halt,* and execute *Right Face.*

d. To obtain *Normal Interval* from *Double Interval,* the command is ***Normal Interval,*** **MARCH.** On the command of execution **MARCH,** the right flank man stands fast. All men to the left of the number one man face to the right as in marching, take one 30-inch step less than their number (for example, number three man takes two steps), *Halt,* and execute *Left Face.*

6-5. ALIGNING THE SQUAD

To align the squad, use the following procedures:

NOTE: The squad leader commands the squad to the appropriate interval before giving the command for alignment.

a. To align the squad at *Normal Interval,* the commands are ***Dress Right,*** **DRESS** and ***Ready,*** **FRONT.** These commands are given only when armed soldiers are at *Order Arms* or *Sling Arms.* On the command of execution **DRESS,** the right flank man stands fast. Each member, except the right flank man, turns his head and eyes to the right and aligns himself with the man on his right. Each member, except the left flank man, extends his left arm laterally at shoulder level, elbow locked, fingers and thumb extended and joined, palm facing down. He ensures his left arm is in line with his body and positions himself by short steps right or left until his right shoulder touches the fingertips of the man on his right. On the command of execution **FRONT,** each member returns sharply to the *Position of Attention* (Figure 6-2).

Figure 6-2. Alignment (Normal Interval).

NOTE: If the squad leader wants exact alignment, on the command of execution **DRESS**, he faces to the *Half Left* in marching and marches by the most direct route to a position on line with the squad, halts one step from the right flank man, and faces down the line. From his position, he verifies the alignment of the squad, directing the men to move forward or backward, as necessary, calling them by name or number: ***"Private Jones, forward 2 inches;" "Number eight, backward 4 inches."*** The squad leader remains at attention, taking short steps to the right or left as necessary to see down the squad. Having aligned the squad, he centers himself on the right flank man by taking short steps left or right. He then faces to the *Half Right* in marching, returns to his position (center of the squad), halts perpendicular to the formation, faces to the left, and commands ***Ready,*** **FRONT**. These procedures also apply when aligning the squad at close or *Double Interval*.

b. To align the squad at *Close Interval,* the commands are ***At Close Interval, Dress Right,*** **DRESS** and ***Ready,*** **FRONT.** The movement is executed in the same manner prescribed for alignment at *Normal Interval* except that the squad members obtain *Close Interval* (Figure 6-3, page 6-6).

Figure 6-3. Alignment (Close Interval).

c. To align the squad at *Double Interval,* the commands are ***At Double Interval, Dress Right,*** **DRESS** and ***Ready,*** **FRONT.** These commands are given only when the troops are unarmed or at *Sling Arms.* On the command of execution **DRESS,** each member (except the right flank man) turns his head and eyes to the right and aligns himself on the man on his right. At the same time, each member (except the right and left flank men) extends both arms and positions himself by short steps right or left until his fingertips are touching the fingertips of the members on his right and left. (The right flank man raises his left arm; the left flank man raises his right arm.)

d. To align the squad in column, the commands are **COVER** and **RECOVER.** On the command **COVER**, each member (except the number one man) raises his left arm to a horizontal position, elbow locked, fingers and thumb extended and joined, palm facing down, and obtains an arm's length plus about 6 inches (from the fingertips) to the back of the man to his front. At the same time, each man aligns himself directly behind the man to his front. To resume the *Position of Attention,* the command **RECOVER** is given. On this command, each member sharply returns to the *Position of Attention.*

6-6. MARCHING THE SQUAD

To march the squad, use the following procedures:

a. For short distances only, the squad may be marched forward while in a line formation.

b. When marching long distances, the squad is marched in column.

c. To form a column formation from a line formation, the command is ***Right,*** **FACE**.

d. When a column formation is originated from a line formation at *Close Interval,* the squad may be marched for short distances at the *Half Step* with less than correct distance. To obtain correct distance while marching with less than correct distance, the command is ***Extend,*** **MARCH**. On the command of execution **MARCH**, the number one man takes one more 15-inch step and then steps off with a 30-inch step. Each squad member begins marching with a 30-inch step at the approximate point where the number one man stepped off, or as soon as correct distance has been obtained.

NOTE: See Chapter 4 for more information on marching movements.

6-7. CHANGING THE DIRECTION OF A COLUMN

To change the direction of a column, use the following procedures:

a. From the *Halt,* the command to start the squad in motion and simultaneously change the direction of march 90 or 45 degrees is ***Column Right (Left),*** **MARCH** or ***Column Half Right (Left),*** **MARCH**. On the command of execution **MARCH**, the lead man faces to the right (left) as in marching by pivoting to the right (left) on the ball of the right foot and steps off in the indicated direction taking a 30-inch step with the left foot and continues to march. The number two man adjusts his step by lengthening or shortening as necessary to reach the approximate pivot point of the lead man. When he reaches the approximate pivot point of the lead man, he pivots to the right (left) on the ball of the lead foot taking a 30-inch step with the trail foot in the new direction. All other members step off with the left foot and continue to march forward taking 30-inch steps and execute in the same manner as the number two man in approximately the same place until the entire squad has executed the column movement.

b. To change the direction of march 90 or 45 degrees when marching, the preparatory command ***Column Right (Left)*** or ***Column Half Right (Half Left)*** is given as the foot (in the desired direction) strikes the marching surface. The command of execution **MARCH** is given the next time the foot in the desired direction strikes the marching surface. On the command of execution **MARCH**, the lead man takes one additional step, pivots in the commanded direction as the pivot foot strikes the marching surface, and continues to march in the new direction. Other members continue to march forward and execute the pivot as prescribed from the *Halt.*

c. To avoid an obstacle in the line of march, the squad leader directs, **INCLINE LEFT(RIGHT).** The lead man inclines left(right) around the obstacle and resumes the original direction. All other members follow the lead man.

6-8. MARCHING TO THE FLANK

The squad may be marched to the flank (for short distances only) when marching in column. The command for this movement is ***Right (Left) Flank,*** **MARCH.** The preparatory command is given as the foot in the desired direction strikes the marching surface, and the command of execution is given the next time the foot in the desired direction strikes the marching surface. On the command of execution **MARCH,** all members take one more step, pivot 90 degrees in the commanded direction on the ball of the lead foot, and step off in the new direction with the trailing foot. As the members begin to march in the new direction, they glance out of the corner of the right eye and dress to the right.

6-9. FORMING A COLUMN OF TWOS AND RE-FORMING

To form a column of twos and re-form, use the following procedures:

a. Forming a column of twos from a file is executed only from the *Halt.* The command is ***Column of Twos to the Right (Left),*** **MARCH.** On the preparatory command, the lead team leader commands **STAND FAST.** The trailing team leader commands ***Column Half Right (Left).*** On the command of execution **MARCH,** the trailing team leader executes a *Column Half Right (Left),* inclines to the left or right when the correct interval is obtained, and commands ***Mark time,*** **MARCH** and ***Team,*** **HALT** to *Halt* abreast of the lead team leader.

b. Forming a file from a column of twos is executed only from the *Halt.* The command is ***File from the Left (Right),*** **MARCH.** On the preparatory command, the lead team leader commands **FORWARD**. The trailing team leader commands **STAND FAST**. On the command of execution **MARCH,** the lead team marches forward. The trailing team leader commands ***Column Half Left (Right)*** when the second man from the rear of the lead team is abreast. He gives the command **MARCH** when the last man of the lead team is abreast of him and his right foot strikes the marching surface. He then inclines right or left to follow the lead team at the correct distance.

NOTE: Commands are given over the team leader's right shoulder if the direction of movement is to the right or if the team is following an element to its right. Commands are given over the left shoulder if the direction of movement is to the left or if the team is following an element to its left.

6-10. DISMISSING THE SQUAD

The squad is dismissed with the members at *Attention.* With armed troops, the commands are ***Inspection,*** **ARMS;** ***Ready, Port,*** **ARMS;** ***Order (Sling),*** **ARMS;** and **DISMISSED.** With unarmed troops, the command is **DISMISSED.**

NOTE: Unless otherwise stated (by the person in charge in his instructions before the command **DISMISSED**), the command **DISMISSED** terminates only the formation, not the duty day (AR 310-25).

Section II. STACK AND TAKE ARMS (M16-SERIES RIFLE)

The squad members execute *Stack Arms* from their positions in line formation (at *Normal Interval*) from *Order Arms*. When in line formation, the squad leader commands ***Count,*** **OFF** and then designates the stack men by numbers (2-5-8).

NOTES:

1. M4-series carbines are not compatible with the M16-series rifles when stacking arms. The two types of weapons must be stacked separately or grounded in a manner that will not damage the sights. (See Appendix D for a detailed explanation of stack arms and take arms for the M4-series carbine.)
2. When the squad is part of a larger unit, stack arms may be executed in a column formation (when the formation consists of three or more files and the squads are at normal interval). Second or third squad is designated as the stack squad.

6-11. PREPARE SLINGS

After the stack men are designated, the squad leader commands ***Prepare,*** **SLINGS.** On the command of execution **SLINGS,** each stack man (or stack squad) grasps the barrel of his rifle with the right hand and raises the rifle vertically. With his left hand, he places the rifle butt on his right hip and cradles the rifle in the crook of his right arm. Using both hands, he adjusts the sling keeper so that a 2-inch loop is formed from the sling keeper to the upper sling swivel. As soon as the loop is prepared, he returns to *Order Arms.*

6-12. STACK ARMS

When all stack men have returned to *Order Arms,* the squad leader commands ***Stack,*** **ARMS.**

a. On the command of execution **ARMS,** each stack man grasps the barrel of his rifle with his right hand and places the rifle directly in front of and centered on his body with the sights to the rear. The rifle butt is placed on the marching surface so that the heel of the rifle butt is on line with the toes of his footgear. The stack man bends slightly forward at the waist and grasps his rifle with his left hand at the upper portion of the handguard (keeping the rifle vertical at all times). The first two fingers of the left hand hold the inner part of the loop against the rifle. The stack man reaches across the front of the rifle with his right hand, grasps the outer part of the loop, and holds it open for insertion of other rifles.

b. On the command of execution **ARMS,** the men to the right and left of the stack man perform the following movements simultaneously:

(1) The man on the stack man's right grasps the barrel of his rifle with his right hand and raises and centers his rifle with the magazine well facing to the front, wrist held shoulder high, elbow locked. With his left hand, he then grasps the handguard (midway), releases his right hand, and regrasps the rifle at the small of the stock. He lowers both arms, with elbows locked (holding the rifle in a horizontal position with the muzzle to the left and the magazine well to the front).

(2) The man on the stack man's left grasps the barrel of the rifle with his right hand and raises and centers his rifle with magazine well facing to the front, wrist held shoulder high, elbow locked. Using his left hand, he then grasps the rifle at the small of the stock, releases the right hand, and regrasps the handguard midway. He then lowers both arms, with elbows locked holding the rifle in a horizontal position with the muzzle to the right and magazine well to the front.

c. As soon as the stack man has placed his rifle in position, both men move the foot nearest the stack man half way *(Half Right* or *Half Left)* toward the stack man. The man on the stack man's left inserts the muzzle of his rifle into the loop to a point about halfway between the flash suppressor and the front sight assembly. He holds his rifle in this position until the man on the stack man's right inserts the muzzle of his rifle in a similar manner and above the other rifle muzzle.

d. Without moving the feet, both riflemen swing the butt of their rifles out and then down to the marching surface, making the stack tight with the rifle butts on line and about 2 feet from the base line. When the stack has been completed, all three men resume the *Position of Attention.*

e. Additional rifles are passed to the nearest stack on the right (right or left if stacked in column). The men with additional rifles grasp the rifle barrel with the right hand and raise the rifle vertically with the magazine well to the front, wrist held shoulder high, elbow locked, and right arm extended to the right front. Throughout the pass, the rifle is held vertical with the magazine well to the front.

(1) The man to the left of the stack man then grasps the rifle midway at the handguard with his left hand. The man passing the additional rifle then releases the rifle and sharply returns to the *Position of Attention.* The man to the left of the stack man then moves the rifle to the right until it is centered on his body, and he grasps the rifle barrel with his right hand, wrist held shoulder high and elbow locked. He then releases the left hand and sharply returns his left hand to the left side as in the *Position of Attention.* He then moves the rifle to his right front.

(2) The stack man receives the rifle and centers it in the same manner as previously described. The man to the left of the stack man sharply returns to the *Position of Attention* after he releases the rifle. Once the stack man has centered the rifle and grasped the barrel with the right hand, he bends forward at the waist and places the rifle in the stack so that it is secure (without damaging the front sight assembly). If there are two additional rifles, the second rifle is passed in the same manner as the first.

NOTE: See Appendix B for a figure showing *Stack Arms* using the M4-series carbine.

6-13. TAKE ARMS

To *Take Arms,* the command is ***Take,*** **ARMS.** On the command of execution **ARMS,** the men return the additional rifles in the same manner as the rifles were received. The stack man secures the stack and holds the loop in the same manner as for stacking rifles. The men on the left and right step toward the stack man in the same manner as when stacking arms. Each man reaches down and regrasps his rifle (one hand at the small of the stock and one hand midway of the hanguard) and brings it to the horizontal position. The man on the right frees his rifle first and resumes *Order Arms.* The man on the left frees his rifle and resumes *Order Arms.* The stack man cradles his rifle and adjusts the sling and sling keeper to its original position and then resumes *Order Arms.*

CHAPTER 7
PLATOON DRILL

"If in training soldiers commands are habitually enforced, the army will be well-disciplined; if not, its discipline will be bad."

Sun Tsu, On the Art of War, trans. 1910

Section I. FORMATIONS

The platoon has two prescribed formations—line and column (Figure 7-1, page 7-2). However, the platoon may be formed into a file or a column of twos from a column formation. When in a line formation, the elements (squads) of a platoon are numbered from front to rear; in a column formation, from left to right. (Appendix G contains an explanation of the symbols used in figures.)

7-1. BASIC INFORMATION

For the most part, platoon drill merely provides the procedures for executing drill movements in conjunction with other squads formed in the same formation. Individual drill movements and the manual of arms are executed as previously described while performing as a squad member during the conduct of platoon drill. For continuity purposes in this chapter, "platoon sergeant" may also denote "platoon leader" when the platoon sergeant is executing drill from his post.

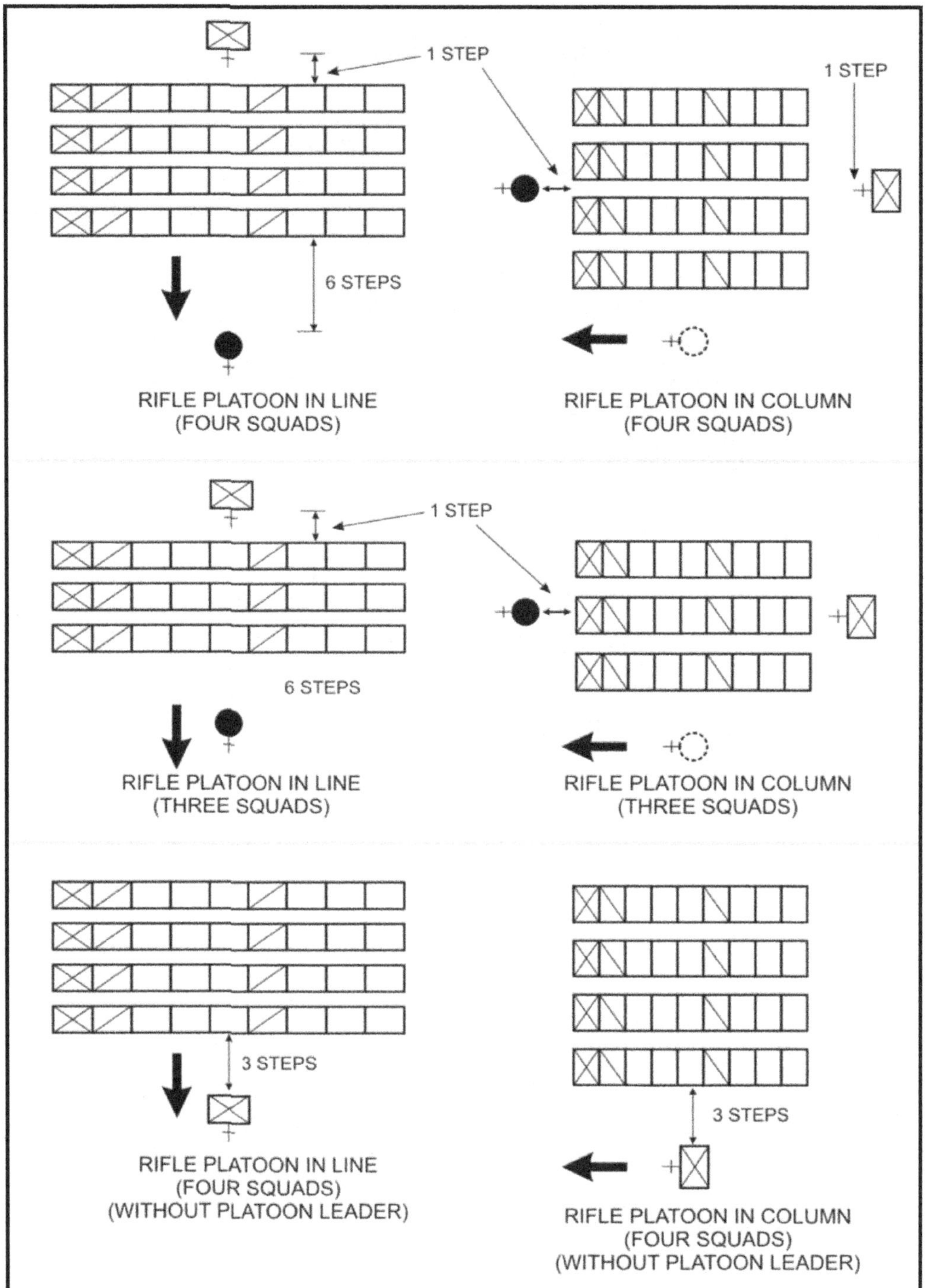

Figure 7-1. Platoon formations.

a. During all drill and ceremonies, the platoon leader and platoon sergeant carry their rifles at *Sling Arms.* When the platoon drills as part of a larger unit, the platoon leader and platoon sergeant remain at *Sling Arms* during all manual of arms movements except when executing the *Hand Salute* while at *Sling Arms.*

b. When the platoon drills as a separate unit or as part of a larger unit in a line formation, without officers present, the post for the platoon sergeant is three steps in front of and centered on the platoon. When in column formation, the post for the platoon sergeant is three steps to left flank of and centered on the platoon.

c. The post for the platoon sergeant with the platoon leader present is one step to the rear and centered on the platoon in line or column formation.

d. When assuming his post in column from a line formation with the platoon leader present, the platoon sergeant faces to the left in marching (on the command of execution **FACE**) and marches in the most direct route to his post. He then halts and faces to the right. When assuming his post in line from a column formation, he faces to the right in marching (on the command of execution **FACE**) and marches in the most direct route to his post, halts centered on the platoon, and faces to the left.

NOTE: When the files (columns) are uneven, the platoon sergeant normally directs the men in longer files to move to another file to balance the formation; or he may fill the vacancy himself as the last man in the right file.

e. When control of the formation is being exchanged between the platoon sergeant and the platoon leader, the platoon sergeant will **always** travel around the **right flank** (squad leader) of the formation when marching from post to post. The platoon leader will **always** travel around the **left flank** of the formation when marching post to post.

f. On the command ***Open Ranks,*** **MARCH;** ***Backward,*** **MARCH**; ***Right (Left) Step,*** **MARCH;** ***Forward,*** **MARCH** and on commands that cause the platoon to change interval in line, he moves at the same time (with the appropriate step) so as to maintain proper position.

g. The leader of the first squad serves as the base when the platoon is a line formation. The leader of the fourth squad serves as the base when the platoon is in a column formation.

h. If for some reason the platoon is authorized a guidon or phase banner (in training units for example), the bearer's post is one step in front of and two 15-inch steps to the right of and facing the person forming the platoon. When the formation is faced to the right for a marching movement, he executes in the same manner as explained in Appendix H, except that his post is three steps in front of and centered on the squad leaders. If the platoon leader is present and at his post, the bearer's post is one step to the rear and two 15-inch steps to the left of the platoon leader.

i. When the platoon drills as a separate unit, in a line formation, the post for the platoon leader is six steps in front of and centered on the platoon; when in a column formation, the platoon leader is six steps to the left flank and centered on the platoon. When marching as part of a larger formation, his post is one step in front of and centered on the squad leaders.

j. When the platoon leader commands ***Open Ranks,*** **MARCH;** ***Backward,*** **MARCH;** ***Right (Left) Step,*** **MARCH;** ***Forward,*** **MARCH** or causes the platoon to change interval, he moves at the same time (with the appropriate step) so as to maintain proper position.

(1) When assuming his post in column from a line formation, the platoon leader faces to the right in marching (on the command of execution **FACE**) and marches in the most direct route to his post, halts, and faces to the left.

(2) When assuming his post in line from a column formation, the platoon leader faces to the left in marching (on the command of execution **FACE**) and marches in the most direct route to his post, halts perpendicular to the formation, and faces to the right.

7-2. FORMING THE PLATOON

The platoon normally forms in a line formation; however, it may re-form in a column when each man can identify his exact position (equipment grounded) in the formation.

a. The platoon forms basically the same as a squad. The platoon sergeant assumes the *Position of Attention* and commands **FALL IN** ***(At Close Interval,*** **FALL IN** or ***In Column,*** **FALL IN).** On the command **FALL IN** ***(At Close Interval,*** **FALL IN),** the squad leader and the first squad (when formed) is three steps in front of and centered on the platoon sergeant. Other squad leaders cover on the first squad leader at the correct distance, which is obtained by estimation. The members of the first squad fall in on their squad leader as prescribed for squad drill. Members of the other squads fall in on their squad leader, assume the *Position of Attention,* and turn their heads and eyes to the right. They obtain correct distance by taking short steps forward or backward and align themselves on the man to their right. They then sharply turn their heads and eyes to the front as in the *Position of Attention* and obtain proper interval by taking short steps left and right to cover on the man to their front. Members of all squads, other than the first squad, will not raise their left arms unless the man to their immediate left has no one to his front in the formation on which to cover.

b. When armed, members fall in at *Order Arms* or *Sling Arms.* For safety, the commands ***Inspection,*** **ARMS;** ***Ready, Port,*** **ARMS;** ***Order (Sling),*** **ARMS** are given at the initial formation of the day and just before the last command, **DISMISSED.**

c. When a report is appropriate, the platoon sergeant commands **REPORT.** The squad leaders, in succession from front to rear, turn their heads and eyes toward the platoon sergeant and salute (holding the *Salute* until returned) and report. The squad leaders do not state the unit. For example:

(1) ***Situation 1.*** When all squad members are in formation, the report is **"All present."**

(2) ***Situation 2.*** When squad members are absent, the soldiers and reasons for absence are reported:

Pvt. Smith - CQ runner.
Pvt. Jones - sick call.
Pvt. Williams –AWOL.

d. The platoon sergeant turns his head and eyes toward the reporting squad leader, receives the report, and returns the salute. After receiving the report from the squad leaders, the platoon sergeant faces about and awaits the arrival of the platoon leader or a directive from the first sergeant to **REPORT**. When the platoon leader has halted at his post, the platoon sergeant salutes and reports, ***"Sir, All present";*** or ***"Sir, All accounted for";*** or ***"Sir, (so many) men absent."*** The platoon leader returns the *Salute.* After the *Salute* has been returned, the platoon sergeant faces to the right in marching, inclines around the squad leaders, halts at his post, and faces to the right. If reporting to the first sergeant, the platoon sergeant turns his head and eyes toward the first sergeant, salutes, and reports.

e. If the platoon leader is not present for the formation, and the commander is in charge of the formation, the platoon sergeant steps forward three steps (after receiving the squad leader's report) and, on the command **POST**, assumes the duties of the platoon leader.

f. When appropriate, the platoon may be formed by the platoon leader rather than by the platoon sergeant. The procedures are the same as previously described except that the first squad forms six steps in front of and centered on the platoon leader, and the platoon sergeant forms at his post to the rear of the platoon. For continuity purposes in this chapter, "platoon sergeant" may also denote "platoon leader" when the platoon leader is executing drill from his post.

7-3. BREAKING RANKS

When the situation requires one or more individuals to leave the formation or to receive specific instructions from the platoon sergeant, the platoon sergeant directs: ***"Private Doe (pause), front and center"****;* or, ***"The following personnel front and center—Private Doe (pause), Private Smith."*** When the individual's name is called, he assumes the position of *attention* and replies, ***"Here, Sergeant (Sir)."*** He then takes one (15-inch) step backward, halts, faces to the right (left) in marching, and exits the formation by marching to the nearest flank. *The soldier does not look left or right.* Once the individual has cleared the formation, he begins to double-time and halts two steps in front of and centered on the platoon sergeant.

NOTE: When a group of individuals are called from the formation, the group forms centered on the platoon sergeant. The platoon sergeant should direct (point) the first man into position so that the rank will be centered when the last man has joined the group.

7-4. COUNTING OFF

The platoon counts off in the same manner as the squad. When in a line formation, the squads count in unison from right to left; each squad leader sounds off, **"ONE."** When in a column formation, the men abreast of each other count in unison from front to rear; each squad leader sounds off, **"ONE."**

7-5. CHANGING INTERVAL

The platoon changes interval in a line formation in the same manner as the squad.

a. To change interval when the platoon is in a column at the *Halt,* the right file stands fast and serves as the base. All other members (abreast of each other) execute the movement as previously described. To obtain *Close Interval* from *Normal Interval,* the third squad takes one right step, the second squad takes two right steps, and the first squad takes three right steps. To obtain *Normal Interval,* the procedures are the same except that the squads execute the same number of left steps.

b. To change interval when the platoon is marching in a column, the preparatory command ***Close Interval*** is given as the right foot strikes the marching surface, and the command of execution **MARCH** is given the next time the right foot strikes the marching surface. On the command of execution **MARCH,** the base squad (right file) takes one more 30-inch step and then executes the *Half Step.* All other men take one more step, simultaneously execute a *Column Half Right,* and march until *Close Interval* is obtained. They execute a *Column Half Left* and assume the *Half Step* when abreast of the corresponding man of the base squad. On the command ***Forward,*** **MARCH,** all men resume marching with a 30-inch step. The commands ***Mark Time,*** **MARCH** and ***Platoon,*** **HALT** may also be given.

c. To resume marching at *Normal Interval,* the preparatory command ***Normal Interval*** is given as the left foot strikes the marching surface, and the command of execution **MARCH** is given the next time the left foot strikes the marching surface. On the command of execution **MARCH,** the platoon members obtain *Normal Interval* in the same manner prescribed for *Close Interval* except that they each execute *Column Half Left* and then *Column Half Right.*

d. To obtain *Double Interval* from *Normal Interval,* the procedures are the same as from *Close Interval* to *Normal Interval.*

e. To obtain *Normal Interval* from *Double Interval* (closing the formation in column from *Open Ranks*), the procedures are the same as obtaining *Close Interval* from *Normal Interval.*

7-6. ALIGNING THE PLATOON

The platoon is aligned similar to the squad.

a. On the command of execution **DRESS,** the first squad leader stands fast and serves as the base. Other squad leaders obtain correct distance by estimation. The members of the first squad execute in the same manner as in squad drill to obtain exact interval. All other squads execute as the first squad, except that each squad member raises the left arm only for uniformity, actually covering (glancing out the corner of the left eye) on the man to the front.

b. If the platoon sergeant wants exact alignment, he faces to the *Half Left (Half Right)* in marching and marches (on the command of execution **DRESS**) by the most direct route to a position on line with the first squad, halts one step from the squad leader or left flank man, and faces down the line. From this position, he verifies the alignment of the first squad, directing the men to move forward or backward, as necessary, calling them by name or number: ***"Private Jones, forward 2 inches; Number eight, backward 4 inches."*** The platoon sergeant remains at *Attention,* taking short steps to the right or left as necessary, to see down the squad.

(1) Having aligned the first squad, the platoon sergeant, after centering himself on the first squad, faces to the left (right) in marching, taking two (three if at open ranks) short steps to the next squad, halts, faces down the line, and aligns the squad in the same manner.

(2) After the last squad is aligned, the platoon sergeant centers himself on the squad leader by taking short steps left or right, faces to the right (left) in marching, returns to his position (centered on the platoon), halts perpendicular to the formation, faces to the left (right), and commands ***Ready,*** **FRONT**.

NOTE: When the platoon dresses as part of a larger formation, all squad leaders turn their heads and eyes to their right and align themselves on the unit to their right. The platoon sergeant aligns the platoon from the left flank rather than from the right flank when his platoon is not formed as the right flank platoon.

c. To align the platoon in column, the commands are **COVER** and **RECOVER**. On the command **COVER**, the fourth squad leader stands fast and serves as base. The squad leaders with the exception of the left flank squad leader, raise their arms laterally and turn their heads and eyes to the right. The members of the fourth squad raise their left arms horizontally (as in squad drill) to the front and cover the man to their front at correct distance (one arm's length plus 6 inches). Squad members of the third, second, and first squads raise their left arms horizontally to the front (for uniformity only), cover on the man to their front, and, at the same time, glance out of the corner of their right eyes aligning on the man to their right. To resume the *Position of Attention*, the command is **RECOVER**. On this command, each man sharply returns to the *Position of Attention*.

NOTE: When the platoon is aligned in column as part of a larger unit but is not the lead platoon, the squad leaders do not raise their arms or turn their heads and eyes but cover on the squad to their front. The platoon sergeant must wait until the platoon to his front has been given the command **RECOVER** before giving the command **COVER.** If the platoon leader is at his post, he must wait until the platoon leader of the platoon to his front has commanded **RECOVER** before he faces about and gives the command **COVER**.

7-7. OPENING AND CLOSING RANKS

To open or close ranks, use the following procedures:

a. ***Open Ranks,*** **MARCH** is executed from a line formation while at the halt. It may be executed while at any of the prescribed intervals. The command for this movement is ***Open Ranks,*** **MARCH**. On the command of execution **MARCH**, the front rank takes two steps forward, the second rank takes one step forward, the third rank stands fast, and the fourth rank takes two steps backward. If additional ranks are present, the fifth rank takes four steps backward, and the sixth rank takes six steps backward.

NOTE: After taking the prescribed steps, the men do not raise their arms. If the platoon sergeant wants exact interval or alignment, he commands ***At Close Interval (At Double Interval)***, ***Dress Right,*** **DRESS**. If the platoon is to align on an element to the right, the squad leaders turn their head and eyes to the right and align themselves with that element.

b. To *Close Ranks,* the command is ***Close Ranks,*** **MARCH**. On the command of execution **MARCH**, the first rank takes four steps backward, the second rank takes two steps backward, the third rank stands fast, and the fourth rank takes one step forward. On the command of execution **MARCH,** the platoon leader and platoon sergeant take the appropriate number of steps to maintain their posts.

7-8. RESTING THE PLATOON

The platoon rests in the same manner prescribed for the squad.

7-9. DISMISSING THE PLATOON

The procedures for dismissing the platoon are basically the same as prescribed for the squad.

a. If the platoon sergeant so desires, the squads may be released to the control of the squad leaders. The platoon sergeant commands **TAKE CHARGE OF YOUR SQUADS,** and *Salutes* are exchanged. The platoon sergeant is no longer part of the formation.

NOTE: Without leaving their positions, the squad leaders turn their heads and eyes over the left shoulder and command **FALL OUT**. The squad leaders move to a position in the immediate area, and command **FALL IN**. Then, they carry out the previous instructions and or give any instructions for actions before the next duty formation.

b. At times the platoon sergeant may want to turn the formation over to a subordinate; for example, to have the subordinate march the platoon to a different location.

(1) The platoon sergeant calls on an individual and the individual breaks rank (paragraph 7-3). The designated soldier posts two steps from and facing the platoon sergeant. The platoon sergeant passes on any additional instructions and *Salutes* are exchanged. After exchanging *Salutes*, the platoon sergeant moves to a position where he can observe, or he may resume his post at the rear of the formation.

(2) The subordinate steps forward two steps, and assumes the post of the platoon sergeant. The platoon sergeant is no longer part of the formation; however, he may march along side of, or behind, the formation to make any corrections or to give further guidance, if necessary.

c. If the platoon leader is at his post, he commands **PLATOON SERGEANT.** The platoon sergeant faces to the left in marching and inclines around the squad leaders' left flank, halts three steps in front of and centered on the platoon, and faces to the right. The platoon leader then commands **TAKE CHARGE OF THE PLATOON,** and *Salutes* are exchanged. The platoon leader is no longer part of the formation. The platoon sergeant remains at his post, faces about, and carries out the platoon leader's instructions.

Section II. MOVEMENTS

The platoon marches in the same manner prescribed for the squad. When the platoon is marching in a line, the first squad leader serves as the guide, and when the platoon is marching in a column, the fourth squad leader serves as the guide. When marching in line, each member of each squad (glancing out of the corner of his right eye) maintains alignment on the man to his right. When marching in column, each member of the first, second, and third squads maintains alignment on the man to his right.

7-10. EYES RIGHT

The platoon renders courtesy during ceremonies or when marching past the Colors by executing *Eyes Right.* The commands for this movement are ***Eyes,*** **RIGHT** and ***Ready,*** **FRONT.**

a. At the *Halt,* all men (on the command of execution, **RIGHT**), while keeping their shoulders parallel to the front, turn their heads and eyes to the right at a 45-degree angle. They focus on and follow the person passing to the front until they are again looking forward, at which time their heads and eyes remain fixed to the front. Only the platoon leader renders the *Hand Salute.*

b. When marching, the preparatory command ***Eyes*** is given as the right foot strikes the marching surface and the command of execution **RIGHT** is given the next time the right foot strikes the marching surface. On the command of execution **RIGHT,** all men, except the right file, turn their heads and eyes to the right and align themselves on the right file while continuing to march. Men in the right file do not turn their heads and eyes but continue looking straight to the front and maintain correct distance. Only the platoon leader salutes. To terminate the courtesy, the preparatory command ***Ready*** is given as the left foot strikes the marching surface, and the command of execution **FRONT** is given the next time the left foot strikes the marching surface. On the command of execution, the men turn their heads and eyes sharply to the front and the platoon leader terminates the *Hand Salute.*

7-11. CHANGING THE DIRECTION OF A COLUMN

The platoon changes the direction of marching basically the same as the squad.

a. During a column movement, the base element is the squad on the flank in the direction of the turn.

b. To change the direction 90 degrees, the command is ***Column Right (Left),*** **MARCH.** On the command of execution **MARCH,** the base squad executes the movement as in squad drill except that the squad leader takes one 30-inch step and then takes up the *Half Step.* The squad leader continues marching with the *Half Step* until the other squad leaders come abreast. The other squad leaders, while maintaining correct (offset) interval, execute a 45-degree pivot and continue marching in an are. As they come on line (abreast) with the base squad leader, they take up the *Half Step.* When all squad leaders are abreast, they step off with a 30-inch step without command. All other platoon members march forward on the command of execution and execute the column movement at approximately the same location as their squad leaders and in the same manner.

NOTE: When the platoon sergeant is marching his platoon as a separate unit or when the company is marching without officers in charge, the platoon sergeant stays centered on his platoon. When the platoon leader and platoon sergeant are marching at their post in column as part of a larger formation, they execute a *Column Half Right (Left)* rather than a 90-degree column movement. After executing the *Column Half Right (Left),* they continue marching in an arc, incline as necessary, and resume their correct positions. After sensing that the squad leaders are abreast of each other, the platoon leader resumes the 30-inch step. At times, the platoon leader may find it necessary to shorten or lengthen his step in order to maintain correct distance from the unit to his front.

c. To change the direction 45 degrees, the command is ***Column Half Right (Left),*** **MARCH.** On the command of execution **MARCH,** the platoon executes the movement in the same manner as a 90-degree turn except that the base squad leader, as well as the other squad leaders, execute a *Column Half Right (Left).*

NOTE: When executing a column movement at *Double Time,* elements adjust the length of their steps so that interval and distance are maintained through and beyond the pivot point.

d. The platoon marches in the opposite direction ***(Rear,* MARCH)** in the same manner as the squad.

e The platoon inclines in the same manner as the squad. The squad nearest the direction of the turn serves as the base. To avoid an obstacle in the path of the march, the platoon leader directs **INCLINE AROUND LEFT (RIGHT).**

f. When space is limited and the platoon sergeant wants to march his unit in the opposite direction (reverse), with the squad leaders at the head of their squads, he commands ***Counter Column,* MARCH.** On the command of execution **MARCH** (at the *Halt),* the first squad marches forward three steps, executes a *Column Right,* marches across the front of the platoon, and executes another *Column Right* just beyond the fourth squad. The second squad steps forward one step, executes a *Column Right,* marches forward, and execute another *Column Right* between the third and fourth squads. The third squad executes two short *Column Lefts* from the *Halt* and marches between the remainder of the third squad and the second squad. The fourth squad marches forward two steps, executes a *Column Left,* marches across the front of the platoon, and executes another *Column Left* between the first and second squads (Figure 7-2).

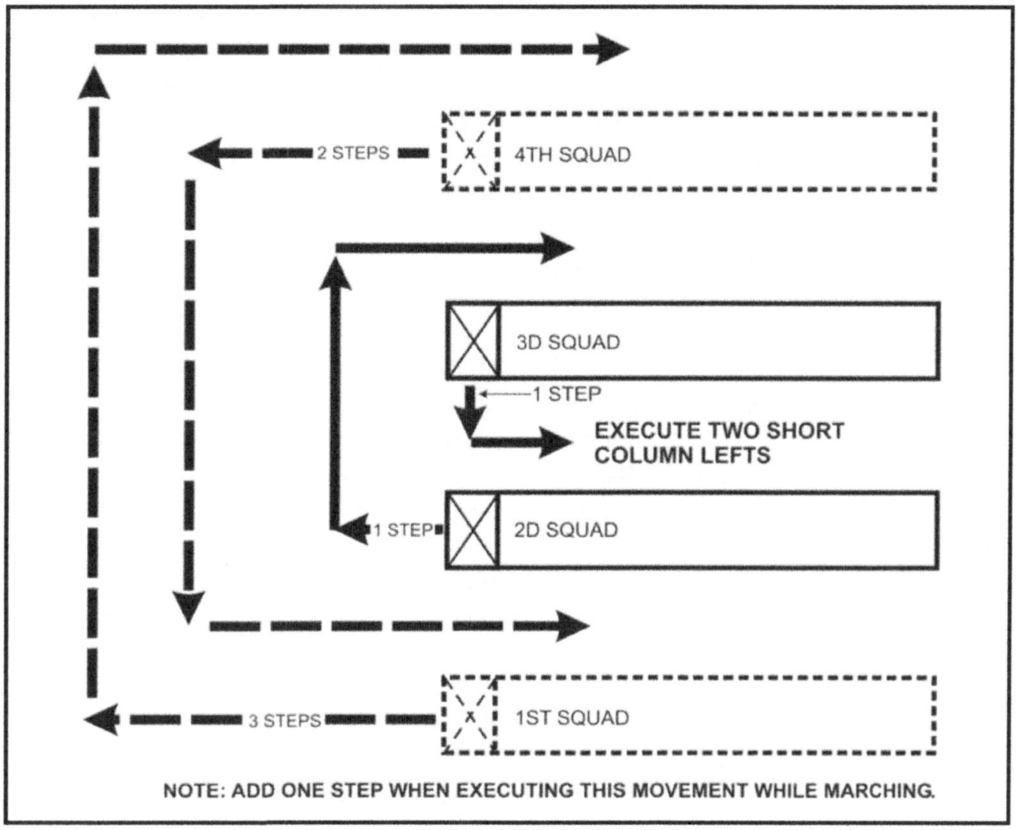

Figure 7-2. Counter-Column March at the Halt.

g. As the third squad leader marches past the last man in the third squad, he and his squad begin to march at the *Half Step.* After marching past the last man in each file, all other squads incline to the right and left as necessary, obtain *Normal Interval* on the third squad, and begin to march with the *Half Step.* When all squads are abreast of each other, they begin marching with a 30-inch step without command.

h. During the movement, *without the platoon leader present,* the platoon sergeant marches alongside the first squad. *With the platoon leader present,* the platoon sergeant marches one step to the rear and centered between the second and third squads (Figure 7-1, page 7-2) and the platoon leader marches alongside the first squad.

i. When marching, the preparatory command ***Counter-Column*** is given as the left foot strikes the marching surface and the command of execution **MARCH** is given the next time the left foot strikes the marching surface. On the command of execution **MARCH**, the platoon executes the movement basically the same as from the *Halt,* except that the squad takes one additional step to ensure that the pivot foot is in the correct position to execute the movement.

NOTES:

1. When there are only three squads in the column, the first squad takes two steps before executing.
2. When the platoon leader and platoon sergeant are marching at their posts as part of a larger formation, the platoon leader takes three steps forward

and executes a column right, marches across the front of the platoon, executes another column right (just beyond fourth squad), marches to his post in the most direct manner, takes up the *Half Step,* and then steps off with a 30-inch step when the squad leaders come on line. The platoon sergeant inclines to the right, follows the third squad until the movement is completed, and then resumes his post.

7-12. MARCHING TO THE FLANKS

The platoon marches to the flank in the same manner as the squad.

7-13. FORMING A FILE AND RE-FORMING

The platoon forms a single file, from the right, left, or as designated, only when in column and at the *Halt.* The commands are ***File from the left (right),*** **MARCH** or ***File in sequence three-two-four-one,*** **MARCH**.

NOTE: If the squad's direction of travel is to the left, or the squad is immediately following an element on the left, then ***all*** supplementary commands will be given over the left shoulder. If the squad's direction of travel is forward or to the right, or is immediately following a squad on the right, ***all*** supplementary commands will be given over the right shoulder. The squad leaders give the appropriate command so that they are following the last soldier in the preceding squad at the correct distance.

a. On the preparatory command, the squad leader of the designated squad gives the supplementary command ***Forward.*** The other squad leaders command **STAND FAST**. On the command of execution **MARCH** (given by the platoon sergeant), the lead squad marches forward. The squad leader next to the lead element (or as designated) looks over the shoulder nearest the moving element and commands ***Column Half Left (Right)*** when the second from the last man is abreast of him and his right foot strikes the marching surface. The *squad leader* gives the command of execution **MARCH** as the right foot of the last man strikes the marching surface when abreast of him. On the command of execution, the squad leader executes the *Column Half Left (Right),* then inclines without command to the right (left), and follows the last man of the preceding squad at correct distance. Other members of the squad march forward and execute the same movements as the squad leader. The remaining squads form the file in the same manner as the squad immediately behind the lead element.

b. The platoon may form a file and execute a column movement simultaneously from a column formation. The command for this movement is ***File from the left (right), Column Left (Right),*** **MARCH**. Following the platoon sergeant's preparatory command, the squad leader of the lead element commands ***Column Left (Right),*** instead of *Forward.* The other squad leaders command **STAND FAST**. On the platoon leader's command of execution **MARCH**, the lead squad executes the *Column Left (Right).* The other squad leaders command ***Column Left (Right)*** at the appropriate time.

NOTES: 1. At this point the platoon is executing platoon drill. The platoon sergeant will not release control of the squads to the squad leaders. The platoon sergeant will rest personnel on an as needed basis; for example, when filing into a building.

2. The squad leaders give the command of execution **MARCH** as the right foot of the last man strikes the marching surface at the pivot *(Column Left),* or the first time the last man's right foot strikes the marching surface after the pivot *(Column Right).*

c. The platoon re-forms to the original column formation only from the *Halt.* The command is ***Column of fours to the right (left),*** **MARCH.**

NOTE: To re-form the platoon to its original column formation when the file was formed on the command ***In sequence __________,*** **MARCH**, the commands **FALL OUT** and **FALL IN** must be given.

(1) On the preparatory command, the squad leader of the base squad commands **STAND FAST.** All other squad leaders command ***Column Half Right (Left).*** On the platoon sergeant's command of execution **MARCH,** the base squad stands fast; all other squads execute the *Column Half Right (Left)* simultaneously.

(2) As each of the moving squad leaders reach a point (line) that ensures correct interval on the element to their left (right), they automatically incline to the left (right) and command ***Mark Time,*** **MARCH** and ***Squad,*** **HALT** so that their squad is abreast of the base (other) squad when halted.

7-14. FORMING A COLUMN OF TWOS AND RE-FORMING

The platoon forms a column of twos from the right or left when in a column of fours at the *Halt.* The command is ***Column of twos from the left (right),*** **MARCH.**

a. On the preparatory command, the squad leaders of the two lead squads command ***Forward.*** The other squad leaders command **STAND FAST.**

(1) On the platoon sergeant's command of execution **MARCH,** the two lead squads march forward. The squad leader (second or third) next to the lead element gives the command to start both of the remaining squads in motion. Looking over the shoulder nearest the moving elements, the squad leader gives the preparatory command ***Column Half Left (Right)*** (when the second from the last man is abreast of him) and then gives the command of execution **MARCH** as the right foot of the last man strikes the marching surface abreast of him.

(2) On the platoon sergeant's command of execution, both squad leaders execute the *Column Half Left (Right),* incline to the right (left) without command, and follow the last men of the preceding squads at correct distance (do not close the space if the files of the two leading squads are not even). Other members of the remaining squads march forward and execute the same movements as their squad leaders.

b. The platoon may form a column of twos and execute a column movement simultaneously from a column formation. The command for this movement is ***Column of twos from the left (right), Column left (right),*** **MARCH.**

(1) The squad leaders of the lead squads command ***Column Left (Right)*** instead of ***Forward.*** The squad leader of the second (third) squad gives the command ***Column Left (Right);*** however, he executes *Column Half Left (Right).* The remaining two squad leaders command **STAND FAST.**

(2) On the platoon sergeant's command of execution **MARCH,** the lead squads execute the *Column Left (Right).* The squad leader (second or third) next to the lead elements gives the command to start both squads in motion. Looking over the shoulder nearest the moving elements, the squad leader gives the preparatory command ***Column Left (Right)*** when the second from the last man is abreast of him. Then he gives the command of execution **MARCH** as the right foot of the last man strikes the marching surface at the pivot *(Column Left)* or the first time the last man's right foot strikes the marching surface after the pivot *(Column Right).* Although the command is ***Column Left (Right),*** the outside squads (fourth or first) execute *Column Half Left (Right).*

c. The platoon re-forms to the original column formation only from the *Halt.* The command is ***Column of fours to the right (left),*** **MARCH.**

(1) When re-forming into a column of fours *without the platoon leader present*, the platoon sergeant remains at his post centered on the platoon. When re-forming into a column of fours *with the platoon leader present*, the platoon leader faces as in marching and resumes his original position in the column. The platoon sergeant follows the rear element and resumes his post in the formation when the element has halted.

(2) On the preparatory command, the squad leaders of the base squads command **STAND FAST.** The trailing squad leaders command ***Column Half Right (Left).*** On the platoon sergeant's command of execution **MARCH,** the base squads stand fast. The trailing squad leaders execute slightly more than the *Column Half Right (Left)* simultaneously. As the trailing squad leaders reach a point (line) that will ensure correct interval on the element to their left (right), they automatically incline to the left (right). The squad leader nearest the stationary squads (second or third) commands ***Mark Time,*** **MARCH** and ***Squads,*** **HALT.**

Section III. INSPECTIONS

Inspections at platoon level are typically conducted by the platoon sergeant. If the platoon leader wants to inspect, he conducts the inspection from his post. For continuity purposes, this section will address the platoon sergeant's actions only.

7-15. BASIC INFORMATION

Consider the following when conducting inspections.

a. The platoon has one prescribed formation for inspecting personnel and equipment in ranks—the line formation. When inspecting crew-served weapons and vehicles, the personnel are normally positioned to the rear of the formation with the operator (gunner) standing by his vehicle (weapon).

b. The platoon leader or platoon sergeant may conduct an in-quarters (barracks) inspection to include personal appearance, individual weapons, field equipment, displays, maintenance, and sanitary conditions. The platoon leader or platoon sergeant will follow local guidelines.

c. When field equipment is to be inspected, it should be displayed as shown in Chapter 8, Figure 8-5 (as a guide). Additional equipment not shown or different models of the equipment should be arranged in a uniform manner established by the local commander. The squad leaders may carry and use materials to record deficiencies if the local chain of command authorizes it.

7-16. IN-RANKS INSPECTION

To conduct in-ranks inspections, use the following procedures:

a. With the platoon in line formation, the platoon sergeant commands ***Count,*** **OFF.** On the command of execution **OFF**, all personnel with the exception of the right flank personnel turn their head and eyes to the right and the right flank personnel count off with "**ONE.**" After the right flank soldiers have counted their number, the soldiers to their left count off with the next higher number and simultaneously turn their head and eyes to the front. All other members of the formation count off in the same manner until the entire formation has counted off.

b. After the platoon has counted off, the platoon sergeant commands ***Open Ranks,*** **MARCH**. On the command of execution **MARCH**, the front rank takes two steps forward, the second rank takes one step forward, the third rank stands fast, and the fourth rank takes two steps backward. If additional ranks are present, the fifth rank takes four steps backward, and the sixth rank takes six steps backward.

NOTE: After taking the prescribed number of steps, the men do not raise their arms. If the platoon leader wants exact interval or alignment, he commands ***At Close Interval (At Double Interval)***, ***Dress Right,*** **DRESS**. (See paragraph 7-6 for more information on aligning the platoon.)

c. At this point, the platoon is ready to be inspected. Typically, the squads are inspected by the squad leaders; however they may be inspected by the platoon sergeant or platoon leader. (See paragraphs 7-17 and 7-18.)

7-17. SQUAD LEADERS' INSPECTION

If the platoon sergeant wants the squad leaders to inspect their squads, he will direct **INSPECT YOUR SQUADS.** *Salutes* are not exchanged.

a. The squad leader marches forward and to the left, inclines as necessary until he is at a point 15 inches in front of and centered on the first man.

NOTE: If the members of the platoon are armed, the squad leaders will sling their weapons diagonally across the back with the muzzle down and to the right. This movement will be executed without command and prior to the squad leader stepping off. (For more information on how to inspect personnel with weapons and how to manipulate the weapon, see paragraph 7-17, c-f.)

b. The squad leader remains at a modified *Position of Attention* moving his head and eyes only. After inspecting at the center position, he takes a short step forward and to the left and inspects, returns to the center and steps forward and to the right and inspects, and returns to the center position.

c. Having inspected the first man, the squad leader faces to the right as in marching and takes one (two if at normal interval) step, halts, and faces the next man at the appropriate distance. The squad leader conducts the inspection for the rest of the soldiers in the squad.

d. After inspecting the last soldier in the squad, the squad leader faces to the right as in marching and marches around behind the squad, inclining as necessary. While the squad leader marches back to his post, he inspects the squad from the rear.

e. After resuming his post, the squad leader turns his head and eyes over his left shoulder and commands his squad to assume *At Ease.*

f. The platoon sergeant remains at his post (inspects the guidon bearer if appropriate). After the last squad has been inspected and is at *At Ease*, the platoon sergeant commands the platoon to *Attention.*

g. After commanding the platoon to *Attention,* the platoon sergeant commands ***Close Ranks,*** **MARCH.** On the command of execution **MARCH**, the first rank takes four steps backward, the second rank takes two steps backward, the third rank stands fast, and the fourth rank takes one step forward. On the command of execution **MARCH,** the platoon leader and platoon sergeant take the appropriate number of steps to maintain their posts.

h. If the platoon is being inspected as part of a larger formation and control of the platoon has not been turned over to the platoon sergeant, he faces about, executes *At Ease*, and awaits further instructions from the first sergeant.

7-18. PLATOON SERGEANT'S/PLATOON LEADER'S INSPECTION

If the platoon sergeant is not going to inspect the entire platoon, he directs the squad leaders of the appropriate squads to inspect their squads. All others will be inspected by the platoon sergeant. When armed, the platoon sergeant slings his weapon in the same manner as the squad leaders.

a. The platoon sergeant faces to the *Half Left* as in marching and marches by the most direct route to a point 15 inches in front of and centered on the first squad leader (or the squad leader of the squad to be inspected). As soon as the platoon sergeant halts in front of the squad leader, he commands the other squads to *At Ease* and inspects the squad leader.

b. The platoon sergeant remains at a modified *Position of Attention* moving his head and eyes only. After inspecting at the center position, he takes a short step forward and to the left and inspects, returns to the center and steps forward and to the right and inspects, and returns to the center position.

c. Having inspected the squad leader, the platoon sergeant faces to the right as in marching and takes one (two if at normal interval) steps, halts, and faces the next man at the appropriate distance. After the platoon sergeant steps off, the squad leader takes a half step forward and faces about. When moving from man to man, the squad leader and platoon sergeant move simultaneously.

d. Having inspected the last soldier in the squad, the platoon sergeant faces to the right as in marching and marches around behind the squad, inclining as necessary, and inspects the squad from the rear.

e. As the platoon sergeant begins to inspect the first squad from the rear, he commands the next squad to *Attention.* The squad leader returns to his post. After the platoon sergeant arrives in front of the next squad leader, he commands the first squad to *At Ease* over the right shoulder.

f. The platoon sergeant and squad leader execute in the same manner as in inspecting the first squad until the entire platoon has been inspected. After inspecting the rear of the last squad, the platoon sergeant marches by the most direct route to his post, halts, faces to the left and commands the platoon to *Attention.*

g. After commanding the platoon to *Attention,* the platoon sergeant commands ***Close Ranks,*** **MARCH.** On the command of execution **MARCH**, the first rank takes four steps backward, the second rank takes two steps backward, the third rank stands fast, and the fourth rank takes one step forward. On the command of execution **MARCH,** the platoon leader and platoon sergeant take the appropriate number of steps to maintain their posts.

h. If the platoon is being inspected as part of a larger formation and control of the platoon has been not been turned over to the platoon sergeant, he faces about, executes *At Ease*, and awaits further instructions from the first sergeant.

7-19. IN-QUARTERS (STAND BY) INSPECTION

When the platoon sergeant wants to conduct an in-quarters inspection, he uses the following procedures:

a. The squad members are positioned on line with their equipment or as near it as possible. The squad leader positions himself in the path of the inspector at a point near his area of responsibility. At the approach of the inspector, the squad leader commands, ***Squad,*** **ATTENTION,** and reports, ***"Sergeant, the platoon is prepared for inspection."***

NOTE: When the situation dictates, the squad leader may report to the inspector outside of the quarters. A designated individual commands the members to **ATTENTION** as the inspector enters the area.

b. The squad leader then guides the inspector along a route dictated by the physical arrangement of the personnel and equipment. Upon entering the area, the highest ranking soldier present commands **AT EASE**. As the inspector approaches each individual or his equipment, the individual automatically assumes the *Position of Attention*. When the inspector has moved to the next man, the last man inspected resumes the position of *At Ease.* As the inspector begins to exit the area after completing the inspection, **CARRY ON** is commanded.

NOTE: When inspecting soldiers' rooms, *Attention* is not commanded. The individuals automatically assume the *Position of Attention* as the inspector enters the room. The inspector then commands **AT EASE**. As the inspector approaches each soldier, the soldier assumes the *Position of Attention* and resumes *At Ease* after he has been inspected.

CHAPTER 8
COMPANY DRILL

"The fundamentals of drill are established daily ... If these maneuvers are all accurately observed and practiced every day then the army will remain virtually undefeatable and always awe inspiring..."

Frederick the Great: "History of My Own Times," trans. 1789

Section I. PROCEDURES AND PERSONNEL

The procedures used and personnel required to conduct company drill are described in this section. (Appendix G contains an explanation of the symbols used in figures.)

8-1. BASIC INFORMATION

This paragraph discusses basic information that applies to conducting company drill.

a. Individual drill movements, manual of arms, and squad and platoon drills are executed as previously described while conducting company drill.

b. For the most part, company drill provides the procedures for executing platoon drill in conjunction with other platoons in the same formation.

c. For drill purposes, a company consists of a company headquarters and two or more platoons. The company headquarters personnel are attached to the platoons to equalize platoon strength without interfering with the permanent squad organization; however, the commander may form the headquarters personnel into a separate platoon at either flank of the company. When headquarters personnel form as a separate element, the senior sergeant serves as the platoon sergeant, and the company executive officer serves as the platoon leader.

d. When in a line or a mass formation, the right platoon serves as the base; when in a column formation, the lead platoon serves as the base.

e. The company may be formed by the commander and the platoon leaders or the first sergeant and the platoon sergeants. If the commander is not scheduled to receive or inspect the company, the first sergeant and platoon sergeants remain at their posts and execute company drill from their posts. For continuity purposes, the term "first sergeant" may be used to denote "commander," just as "platoon sergeant" may be used to denote "platoon leader" when the first sergeant and platoon sergeants are conducting company drill from their posts.

f. When the first sergeant directs that the company open or close ranks, align, stack or take arms, extend march, close on the leading platoon, or prepare for inspection, the movements are executed on the command of the platoon sergeants and not on the directives of the first sergeant. The platoon sergeants command the movement in sequence beginning with the base platoon.

g. During all drills and ceremonies, the first sergeant and executive officer carry their weapons at *Sling Arms.* They remain at *Sling Arms* during all manual of arms movements except when they execute the *Hand Salute* while at *Sling Arms.*

h. The members of a company break ranks in the same manner as in platoon drill except that the individuals called from the formation form on the first sergeant rather than on the platoon sergeant.

i. The company marches, rests, and executes *Eyes Right* in the same manner as the platoon.

j. The company has four prescribed formations: company in line with platoons in line; company in column with platoons in column; company in column with platoons in line (used primarily for ceremonies); and company mass formation. However, the company may be formed into a column of twos in the same manner as the platoon.

8-2. POSTS FOR KEY PERSONNEL

Key personnel assume their posts as follows:

a. **Company Commander.** The company commander's post is normally 12 steps in front of the front rank of troops and centered on the company. The only exceptions to this rule are when the company forms as part of a larger unit in a column formation (with the company and platoons in column) or when the company is formed in a mass formation. In these situations, the company commander's post is 6 steps in front of and centered on the front rank of troops and 12 steps behind the last rank of troops of the element to his front.

b. **Guidon Bearer.** The guidon bearer's post is two 15-inch steps to the rear and two 15-inch steps to the left of the company commander (Figure 8-1). (The manual of the guidon is discussed in further detail in Appendix H.)

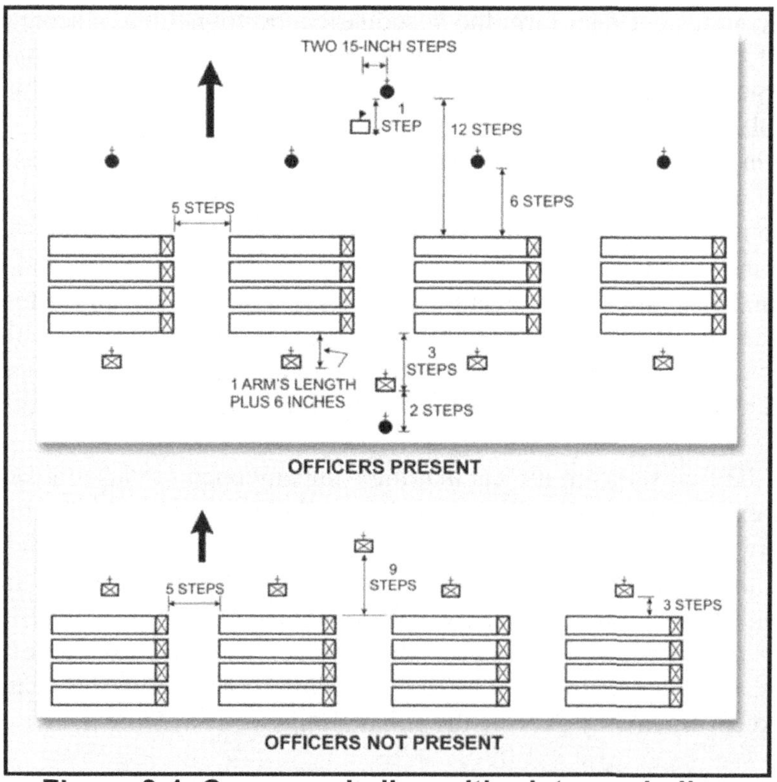

Figure 8-1. Company in line with platoons in line.

NOTE: The company commander should inconspicuously direct the guidon bearer to remain in position and move with the commander when the company is to be faced to the right and marched for only a short distance.

c. **Executive Officer.** When the company is in a line formation, the executive officer's post is two steps to the rear of the first sergeant; in a column or a mass formation, he is three steps to the rear of the last rank and off center (one 15-inch step to the right) of the company.

d. **Platoon Leaders.** When the company is in line formation with the platoons in line, the post for the platoon leader is six steps in front of and centered on his platoon. When the company is in column formation with the platoons in column or the company is in column with the platoons in line, the post for the platoon leader is one step in front of and centered on the platoon. When the company is in mass formation, the post for the platoon leaders is one step in front of and evenly spaced across the company front. If the commander is not present, the platoon leaders are not part of the formation.

e. **First Sergeant.** When the company is in a line formation, the first sergeant's post is three steps to the rear of the last rank and centered on the company; in a column or a mass formation, he is three steps to the rear of the last rank and off center (one 15-inch step to the left) of the company (Figure 8-2, page 8-4).

(1) When the company is in mass formation, *without the commander present*, the post for the first sergeant is five steps in front of and centered on the formation. When the company is in line formation with the platoons in line *without the commander present*, the post for the first sergeant is nine steps in front of and centered on the company. When the company is in column formation with the platoons in column *without the commander present,* the post for the first sergeant is nine steps to the left flank of and centered on the company.

(2) When marching the company in mass formation with the commander at his post, the first sergeant remains at his post to provide cadence and control.

f. **Platoon Sergeant.** When the company is in any of the four formations and the officers are present, the platoon sergeant's post is one step to the rear and centered on his platoon.

(1) When the company is in mass formation, the post for the platoon sergeants is one step to the rear of the company and evenly spaced across the frontage. When the company is in mass formation, *without the officers present*, the post for the platoon sergeants is one step in front of and evenly spaced across the front.

(2) When the company is in line formation with the platoons in line *without the officers present*, the post for the platoon sergeant is three steps in front of and centered on his platoon. When the company is in column formation with the platoons in column *without the officers present,* the post for the platoon sergeant is three steps to the left flank of and centered on his platoon.

(3) When marching the company *without the officers present*, and the company is in column formation with the platoons in column, the platoon sergeants remain at their posts and provide for cadence and control of their platoons.

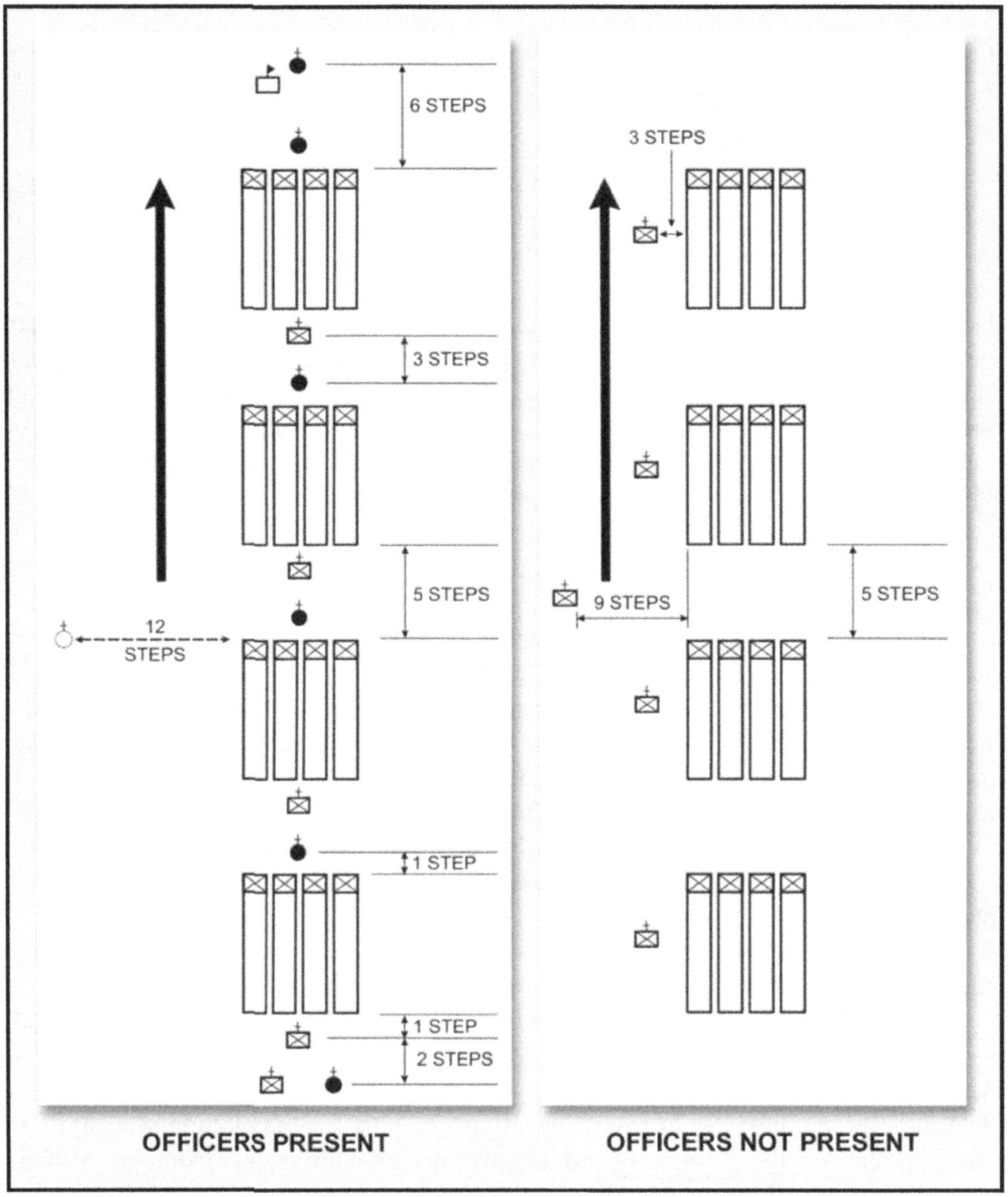

Figure 8-2. Company in column with platoons in column.

NOTE: The executive officer inconspicuously gives the necessary commands (*March, Column Half Left, Halt, and Facings*) for himself and the first sergeant when changing from one post to another. Once in motion, they incline as necessary and march to their new posts. When the company changes from a line formation to column formation, the first sergeant and executive officer face with the company but do not march forward. They remain in position until the last platoon has passed. If the company marches forward (column formation) for a short distance only, or if it closes on the base platoon (line formation) at close interval, the first sergeant and executive officer march forward or face to the right in marching and resume their posts. When the company changes from a column (platoons in column) to a line formation, they face to the right in marching and assume their posts in line. When the company forms a company in column with platoons in line, as the unit is faced, they face to the right in marching and assume their posts to the rear of the last platoon. When the company is formed in mass, they face to the right in marching and assume their posts at the center of the company immediately after halting with the last platoon (Figure 8-3). The off-center position explained for the first sergeant and executive officer should place them at normal interval from each other while covering a file to their front. If either one is not present during a formation, the one present changes posts without command and assumes the post at the center of the company rather than off center (Figure 8-4, page 8-6).

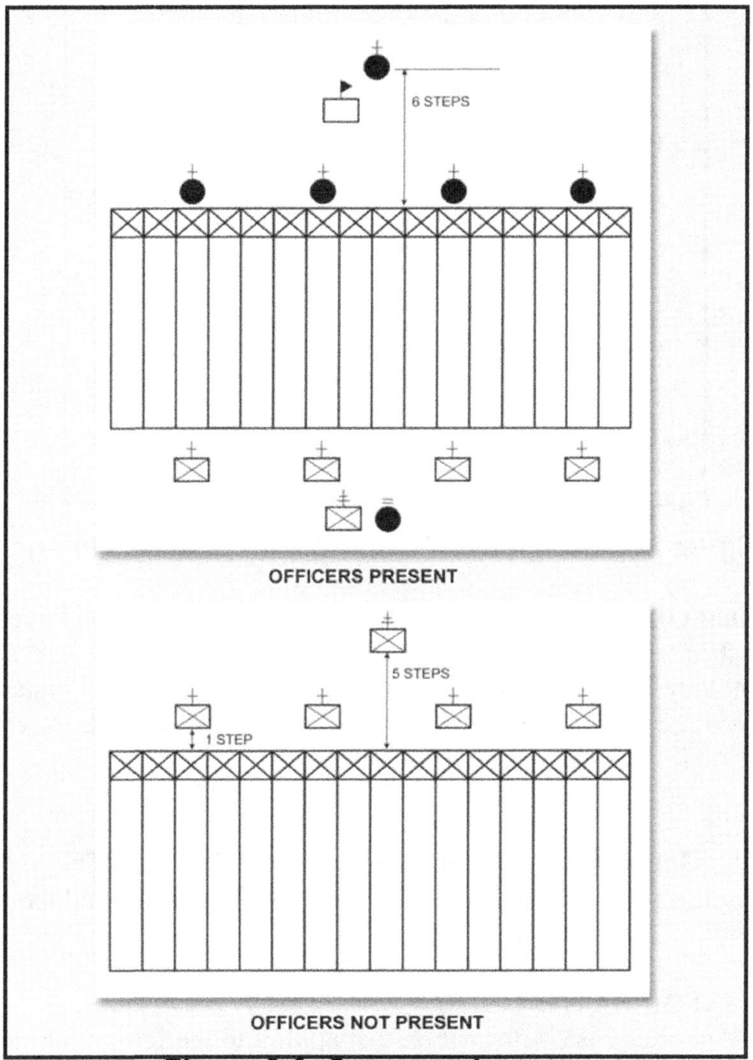

Figure 8-3. Company in mass.

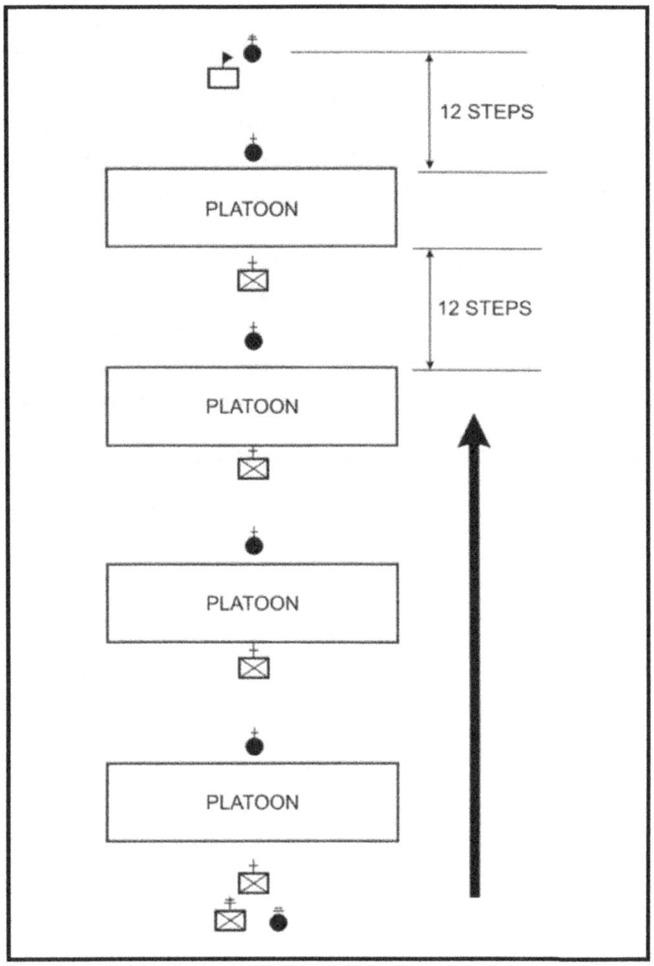

Figure 8-4. Company in column with platoons in line.

g. **Additional Officers.** Other officers assigned or attached, who have no prescribed position within the formation, form evenly to the right and left of the executive officer in a line formation; they fall out and form in one or two ranks at correct distance to the rear of the first sergeant and executive officer when the company forms in column (Figure 8-4).

Section II. FORMATIONS AND MOVEMENTS

This section discusses the formations and movements used in conducting a company drill.

8-3. BASIC INFORMATION

This paragraph discusses basic information that applies to the formations and movements used in conducting a company drill.

a. The company normally forms in a line formation; however, it may re-form in column when each man can identify his exact position (equipment grounded) in the formation.

b. The company forms basically the same as the platoon. On the command **FALL IN;** ***(At Close Interval),*** **FALL IN,** the platoons form in line, centered on and facing the person forming the unit, with five-step intervals between platoons.

NOTE: To have the company assemble in a company mass formation the command is ***Mass Formation,*** **FALL IN.** Before giving the commands the person forming the unit announces the interval and the number of personnel in the front and designates the base man. In this situation, the first sergeant and platoon sergeants travel around the right flank of the formation when moving from post to post. The commander and platoon leaders travel around the left flank of the formation when moving from post to post.

c. When armed, members fall in at *Order (Sling) Arms.* For safety, the person forming the unit commands ***Inspection,*** **ARMS;** ***Ready, Port,*** **ARMS;** ***Order (Sling),*** **ARMS** at the initial formation of the day, or when the last command is **DISMISSED**.

d. The company may be formed by the first sergeant and platoon sergeants or by the company commander and platoon leaders. When possible, the platoons assemble near the formation site before the arrival of the first sergeant or company commander. If the company is formed by the noncommissioned officers, the platoon leaders normally observe the procedures from a position to the rear of their platoons.

(1) When the company is formed by the noncommissioned officers, the following procedures apply.

(a) The first sergeant posts himself nine steps in front of (center) and facing the line where the front rank of each platoon is to form. He then commands **FALL IN;** ***(At Close Interval),*** **FALL IN.**

(b) On the command of execution, the platoons form in the same manner prescribed in platoon drill. Each platoon sergeant faces his platoon while the platoons are forming and directs his platoon to adjust (if necessary) and align on the platoon to its right at the correct interval (as described in paragraph 8-5). Once the platoon is formed, the platoon sergeants face about.

NOTE: If the command ***At Close Interval,*** **FALL IN** is given, the members of the platoon form at close interval; however, the five-step interval between platoons is maintained.

(c) When all of the platoon sergeants are facing to the front, the first sergeant commands (if appropriate) ***Inspection,*** **ARMS;** ***Ready, Port,*** **ARMS;** ***Order (Sling),*** **ARMS.** He then directs (if appropriate) **RECEIVE THE REPORT.** The platoon sergeants face about and command **REPORT.** The squad leaders report as previously described in Chapter 7. Having received the report, the platoon sergeants face about. When all platoon sergeants are facing to the front, the first sergeant commands **REPORT.** The platoon sergeants turn their head and eyes toward the first sergeant, salute and report (as previously described in Chapter 7) in succession from right to left.

The first sergeant turns his head and eyes toward the reporting platoon sergeant and returns each *Salute* individually. Having received the report from the platoon sergeants, the first sergeant faces about and awaits the arrival of the company commander if the commander is scheduled to receive the company.

NOTE: If the commander is not scheduled to receive or inspect the company, the first sergeant and platoon sergeants remain at their posts and execute company drill from their posts. For continuity purposes, the term commander may also be used to denote the first sergeant, just as platoon leader may be used to denote the platoon sergeant when the commander and platoon leaders are conducting company drill from their posts.

(d) When the company commander has halted at his post, the first sergeant salutes and reports, ***"Sir, all present,"*** or ***"Sir, all accounted for,*** "or ***"Sir, (so many) men absent."*** The company commander returns the *Salute* and commands **POST.** The first sergeant faces about and marches to his post three steps to the rear and at the center of the company, halts, and faces about. The guidon bearer steps forward three steps. The platoon sergeants face to the right in marching and assume their posts to the rear of their platoons (if the platoon leader is not present, they step forward three steps). The platoon leaders march around the left flank of their platoons and assume their posts by inclining facing to the front. The company executive officer assumes his post two steps to the rear of the first sergeant.

(2) When the company is formed by the company commander, the procedures are the same as forming with the noncommissioned officers except that the platoon leaders form their platoons and the first sergeant, platoon sergeants, and guidon bearer fall in at their posts (Figure 8-1, page 8-2). The command **POST** is not necessary. If a platoon sergeant is to fill the post of platoon leader, he takes a position six steps in front of and centered on the platoon.

8-4. CHANGING INTERVAL

The company changes interval in the same manner as prescribed for the platoon.

a. When the first sergeant wants the company to obtain close interval in a line formation while maintaining a five-step interval, he directs **CLOSE ON THE BASE PLATOON AT CLOSE INTERVAL.** The platoon sergeants face about and command ***Count,*** **OFF**. After the platoons have counted off, the platoon sergeants command ***Close Interval,*** **MARCH.** The second, third, and fourth platoon sergeants command ***Right,*** **FACE** and in succession command ***Half Step,*** **MARCH.** They halt at the five-step interval and face the platoon to the left. If the directive **CLOSE ON THE THIRD PLATOON AT CLOSE INTERVAL** is given, the platoon sergeants on the right of the designated platoon have their platoons obtain close interval, face their platoons to the left, march *(Half Step)* forward until the five-step interval is obtained, and then halt and face their platoons to the right.

NOTE: When the company is at close interval (line formation), the first sergeant may march the company (in column with less than correct distance) at the *Half Step* for short distances. If he wants to march with a 30-inch step while marching with less than correct distance, he commands ***Extend,*** **MARCH**. The first platoon sergeant commands ***Extend,*** **MARCH**. The other platoon sergeants command **CONTINUE TO MARCH**. As the first platoon begins to march with a 30-inch step, the second platoon sergeant commands ***Extend,*** **MARCH**. The remaining platoons execute the directive in the same manner as the second platoon.

b. When the first sergeant wants the company to obtain normal interval from close interval in a line formation while maintaining a five-step interval, he directs **EXTEND ON THE BASE PLATOON AT NORMAL INTERVAL.** The platoon sergeants face about and march *(Half Step)* their platoons to a position that ensures the five-step interval between platoons after they have obtained normal interval. After halting and facing the platoons to the left, the platoon sergeants command ***Count,*** **OFF**. The platoon sergeants then command ***Normal Interval,*** **MARCH.** If necessary, the platoon sergeant can verify interval as described in the note of paragraph 8-6.

8-5. CHANGING DISTANCE

To increase the distance between elements of a company (or larger unit) while marching in column to normal distance, use the following procedures.

NOTE: This movement is only executed from the half step when a company or larger size element is marching in column at less than normal distance.

a. To increase the distance between elements of a company (or larger unit) while marching in column at the half step to normal distance, the command is ***Extend,*** **MARCH**. On the preparatory command ***Extend*** of ***Extend,*** **MARCH**, the platoon sergeant echoes the preparatory command and all subsequent platoon sergeants issue the supplementary command **CONTINUE TO MARCH**. On the command of execution **MARCH** of ***Extend,*** **MARCH**, the front rank of the lead element takes one more 15-inch step, then steps off with a 30-inch step. All other ranks continue to march with a 15-inch step until the rank to their front has stepped off with a 30-inch step and has obtained normal distance (40 inches).

b. Subsequent platoon sergeants issue the command ***Extend,*** **MARCH** after the entire element to their front has obtained normal distance and has begun marching forward with a 30-inch step.

NOTE: From the *Halt*, the first sergeant directs **HAVE YOUR PLATOONS EXTEND MARCH.** The lead platoon sergeant gives the command ***Extend,*** **MARCH**. All subsequent platoon sergeants give the command ***Half Step,*** **MARCH**. On the command of execution **MARCH** of ***Extend,*** **MARCH**, the front rank steps off with a 30-inch step. All other ranks step off with a 15-inch step and execute the movement in the same manner as previously described.

8-6. ALIGNING THE COMPANY

To align the company, use the following procedures:

a. To align the company in a line formation, the first sergeant directs **HAVE YOUR PLATOONS DRESS RIGHT.**

(1) On the directive, all platoon sergeants face about. The right flank platoon sergeant commands ***Dress Right,*** **DRESS** and aligns his platoon as described in platoon drill. When the alignment of the first rank of the right platoon has been verified (the platoon sergeant has stepped off to align the second rank), the platoon sergeant to the left commands ***Dress Right,*** **DRESS.** He then faces to the half right in marching, moves to a position on line with and one step to the left of the left flank man of the first rank, and faces *(Left Face)* down the line.

(2) After aligning the first rank, the platoon sergeant centers himself on the first rank, faces to the right in marching, takes two short steps, halts, executes *Left Face,* and aligns the second rank. The third and fourth ranks are aligned in the same manner as the second rank.

(3) After aligning the fourth rank, the platoon sergeant faces to the left in marching; returns to his position, center of the platoon; halts perpendicular to the formation; faces to the right; commands ***Ready,*** **FRONT;** and faces about. All platoon sergeants to the left of the second platoon take the same actions as the second platoon sergeant.

NOTE: If necessary, the platoon sergeants to the left of the base platoon command **VERIFY INTERVAL** before commanding ***Dress Right,*** **DRESS**. On the command **VERIFY INTERVAL**, the base (squad leader of the first squad) faces to the right in marching and moves to the left flank member (first squad) of the platoon to the right, halts, faces about, steps forward five steps, halts, faces to the right, and aligns himself on the element to his right.

b. To align the company in column, the first sergeant directs **HAVE YOUR PLATOONS COVER.** On this directive, the first platoon sergeant faces about and commands **COVER.** The other platoon sergeants command **STAND FAST.** The first platoon covers as in platoon drill. The other platoons then execute the movement in succession as soon as the platoon to their front has completed the movement.

8-7. OPENING AND CLOSING RANKS

To open and close ranks, use the following procedures:

a. To *Open Ranks,* the first sergeant directs **HAVE YOUR PLATOONS OPEN RANKS AND DRESS RIGHT.** On the directive, all platoon sergeants face about. The right flank platoon sergeant commands ***Open Ranks,*** **MARCH.** When the platoon has completed the movement, he then commands ***Dress Right,*** **DRESS** and aligns the platoon the same as in platoon drill. When the first rank of the right platoon has been aligned, the platoon sergeant to the left commands ***Open Ranks,*** **MARCH** and then commands ***Dress Right,*** **DRESS.** Then he aligns his platoon in the same manner described in paragraphs 7-6 and 8-5. All platoon sergeants to the left of the second platoon take the same actions as the second platoon sergeant.

NOTE: When clearing grounded equipment, the company may march in column (right face) from open ranks *(Double Interval)* for short distances. Having cleared the equipment, the first sergeant commands ***Normal Interval,*** **MARCH**. When returning to the equipment while marching at normal interval, the command ***Double Interval,*** **MARCH** is given. If the company is marching in reverse order, the commands **FALL OUT** and **FALL IN** should be given.

b. To *Close Ranks,* the first sergeant directs **HAVE YOUR PLATOONS CLOSE RANKS.** On the directive, all platoon sergeants face about and in sequence from right to left command ***Close Ranks,*** **MARCH.** The platoons execute the movement the same as in platoon drill. After the platoons have completed the movement, the platoon sergeants face about.

8-8. CHANGING THE DIRECTION OF MARCH OF A COLUMN

To change the direction of march of a column, use the following procedures.

a. The company changes the direction of march basically the same as the squad and platoon. The commands are ***Column Right (Left), Column Half Right (Left),*** **MARCH.**

b. The base element during a column movement is the lead platoon and the squad on the flank, in the direction of turn.

c. When at the *Halt,* the leading platoon sergeant repeats the first sergeant's preparatory command. Succeeding platoon sergeants give the supplementary command ***Forward.*** On the command of execution **MARCH,** the leading platoon executes the movement as described in platoon drill; succeeding platoons execute the movement on their platoon sergeant's command at approximately the same location.

d. While marching, the movement is executed as described from the *Halt* except that the succeeding platoon sergeants give the supplementary command ***Continue to march*** rather than ***Forward.***

e. The company executes *Rear March* and inclines in the same manner as the platoon.

f. When executing *Counter-Column March* from the *Halt,* the leading platoon sergeant repeats the preparatory command. Succeeding platoon sergeants give the supplementary command ***Forward.***

(1) On the command of execution **MARCH,** the leading platoon executes the movement as described in platoon drill, and marches through the other platoons. Succeeding platoons execute the movement on the platoon sergeant's command at approximately the same location.

(2) When the movement is executed while marching, the command of execution is given as the left foot strikes the marching surface. The movement is executed basically the same as from the *Halt* except that the succeeding platoon sergeants give the supplementary command *Continue to March* rather than *Forward.*

(3) The guidon bearer faces to the left in marching from the *Halt* or executes a *Column Left* in marching, marches by the most direct route outside of the formation, and repositions himself in front of the lead platoon as it clears the rear of the company.

(4) If the first sergeant gives the command from his post in a separate unit, he moves in the most convenient manner to his new position at the left flank of the unit. If he gives

the command while at the head of the company, he moves in the same direction as the guidon bearer.

8-9. CLOSING THE DISTANCE BETWEEN PLATOONS

This movement is only executed when the company is in column with the platoons in column and has been faced to the right for marching with the first sergeant and platoon sergeants at their posts.

a. The first sergeant directs **CLOSE ON THE LEADING PLATOON.** The leading platoon sergeant commands **STAND FAST**. The leading platoon sergeant then faces to the left as in marching and marches, inclining as necessary, to the trail of the company.

b. The succeeding platoon's platoon sergeant marches his platoon forward and commands ***Mark Time,*** **MARCH.** He ensures that he gives the command of execution **MARCH** when the squad leaders are three steps from the last soldier in the leading platoon so that the soldiers of the succeeding platoon begin marching in place at the correct distance.

c. While *Marking Time* in formation, the soldiers adjust position to ensure proper alignment and cover. The proper distance between soldiers while marching is one arm's length plus 6 inches (approximately 36 inches). If necessary, the squad leaders take the appropriate number of steps to close any gaps should the trail of the leading platoon be uneven. The rest of the formation adjusts as necessary.

d. After sensing that the members of the platoon have obtained proper alignment and cover, the platoon sergeant commands ***Platoon,*** **HALT.** After commanding the platoon to *Halt,* the platoon sergeant faces to the left as in marching and marches, inclining as necessary, to the trail of the company.

e. The remaining platoons execute in the same manner as the succeeding platoon in order from front to rear. The platoon sergeants wait until the platoon sergeant to their front gives the command ***Mark Time,*** **MARCH,** before giving the command ***Forward,*** **MARCH.**

f. The platoon sergeants form at the trail of the company one step to the rear and evenly spaced. The platoon sergeants are now part of the formation. If the first sergeant wants exact cover and alignment, he commands **COVER.** The first sergeant remains at his post and marches the company.

8-10. CORRECTING THE DISTANCE BETWEEN PLATOONS

To obtain correct distance (five steps) when the company is marching in column or is in column at the *Halt,* the first sergeant directs **CORRECT ON LEADING PLATOON.**

a. When at the *Halt,* on the directive **CORRECT ON LEADING PLATOON,** the platoon sergeant of the leading platoon commands (over the right shoulder) **STAND FAST.** The succeeding platoon sergeants command (over the right shoulder) ***Forward,*** **MARCH** and then command ***Mark Time,*** **MARCH** and ***Platoon,*** **HALT** when correct distance is obtained.

b. While marching, on the directive **CORRECT ON LEADING PLATOON,** the platoon sergeant of the leading platoon commands (over the right shoulder) ***Half Step,*** **MARCH.** The succeeding platoon sergeants command (over the right shoulder)

CONTINUE TO MARCH, and then command ***Half Step,*** **MARCH** as soon as the correct distance is obtained.

c. The first sergeant commands ***Forward,*** **MARCH; (HALT)** as soon as all platoons have obtained the correct distance and are marching at the half step.

8-11. FORMING A COLUMN OF TWOS AND RE-FORMING

The company forms a column of twos basically the same as the platoon. The first sergeant must allow sufficient time for the platoon sergeants, and the squad leaders of the lead platoon, to give their supplementary commands before giving the command of execution.

a. The command for this movement is ***Column of twos from the right (left),*** **MARCH.** The leading platoon sergeant repeats the preparatory command. Other platoon sergeants give the supplementary command ***Stand Fast.*** On the first sergeant's command of execution **MARCH,** the leading platoon executes the movement as in platoon drill. Other platoons execute the movement on their platoon sergeant's command. Succeeding platoon sergeants give their commands so as to follow with the prescribed five-step distance between platoons.

b. Re-forming into a column of fours is executed only at the *Halt.* The command for this movement is ***Column of fours to the left (right),*** **MARCH.** On the first sergeant's command of execution, all platoons execute the movement simultaneously as described in platoon drill. As soon as the platoons are re-formed, the platoon sergeants automatically march the platoons forward and obtain the five-step distance between platoons.

8-12. FORMING A COMPANY MASS

The company may form in mass from a company in column (platoons in column) when halted or while marching. The company must be at *Close Interval* (***Close Interval,*** **MARCH**) before the command ***Company Mass Left,*** **MARCH** is given.

a. On the preparatory command ***Company Mass Left***, given at the *Halt,* the leading platoon sergeant commands ***Stand Fast.*** The platoon sergeants of the succeeding platoons command ***Column Half Left.*** On the command of execution **MARCH,** the leading platoon *stands fast.* The other platoons execute the *Column Half Left* and then execute a *Column Half Right* on the command of the platoon sergeants to a point (line) that ensures the platoons will be at *Close Interval* alongside the platoon to their right when halted. As the platoons come abreast of the base platoon, the platoon sergeants command ***Mark Time,*** **MARCH.** While the platoon is marking time, the members adjust their positions to ensure alignment on the man to their right. The platoon sergeants allow their platoons to *Mark Time* for about eight counts and then command ***Platoon,*** **HALT**.

b. On the preparatory command ***Company Mass Left*** while marching, given as the left foot strikes the marching surface, the leading platoon sergeant gives the supplementary command ***Mark Time.*** The succeeding platoon sergeants command ***Column Half Left.*** On the command of execution **MARCH,** the leading platoon begins to march in place. After marking time for about eight counts, the lead platoon sergeant commands ***Platoon,*** **HALT.** The other platoons form in the same manner as from the *Halt.*

NOTES: 1. If the commander is in charge of the formation, on the command of execution **MARCH**, the first sergeant and guidon bearer halt and immediately face to the right (left) in marching and reposition themselves centered on the company.

2. If the platoon sergeants are marching at their posts on the left flank of their platoons, they must wait until the platoon sergeants to their front have commanded **HALT** and have moved to the front of the company before moving into position.

8-13. ALIGNING A COMPANY IN MASS

To align a company in mass, use the following procedures:

a. As soon as the company has formed in mass, the first sergeant gives the command ***Order,*** **ARMS** (if appropriate) and then commands ***At Close Interval, Dress Right,*** **DRESS.**

b. On the command of execution **DRESS,** the platoon sergeant of the right platoon marches by the most direct route to the right flank and verifies the alignment of as many ranks as necessary to ensure proper alignment in the same manner as aligning the platoon. When he has finished the verification, the platoon sergeant returns to a position one step in front of and centered on the third squad, halts, and faces to the right. When the platoon sergeant has returned to his position, the first sergeant commands ***Ready,*** **FRONT.** Platoon sergeants align themselves to the right. The first sergeant directs the platoon sergeants to move left or right to ensure the platoon sergeants are evenly spaced across the front.

8-14. CHANGING THE DIRECTION OF MARCH OF A MASS FORMATION

The company changes the direction of march in mass basically the same as a platoon column movement. This movement is normally executed with the commander at his post. For clarity purposes, the term "commander" is used.

a. When executed from the *Halt,* the commander has the unit execute *Right Shoulder Arms* (if appropriate), then faces in the desired direction of march, turns his head toward the formation, and commands ***Right (Left) Turn,*** **MARCH.**

(1) On the command of execution **MARCH,** the platoon leaders face to the half right (left) in marching and continue to march in an arc until parallel to the new direction of march. Then they begin marching with the *Half Step,* dressing on the right (left) flank platoon leader until the command ***Forward,*** **MARCH** is given.

(2) The right (left) guide (the base squad leader in the direction of turn) faces to the right (left) in marching, takes one 30-inch step in the indicated direction, and then takes up the *Half Step.* All other squad leaders (front rank) face to the half right (left) in marching and continue to march in an arc until they come on line with the guide.

(3) At this time, they begin marching with the *Half Step* and dress (glancing out of the corner of the eye) in the direction of the turn until the command ***Forward,*** **MARCH** is given. On that command, the dress is automatically to the right. All other members march forward and execute the movement in the same manner as their squad leaders.

b. When executed while marching, the movement is the same as from the *Halt* except that the company commander faces about (marching backward) to give the command ***Right (Left) Turn,*** **MARCH.** He then faces about and completes the turning movement himself. After the company has completed the turn, he faces about, commands ***Forward,*** **MARCH,** and again faces about.

8-15. FORMING A COLUMN FROM A COMPANY MASS

This movement is normally executed with the commander at his post. For clarity purposes, the term "commander" is used. To form a company in column from a company mass at the *Halt,* the command is ***Column of Platoons, Right Platoon, Column Right (Column Half Right),*** **MARCH.** The right platoon leader gives the supplementary command of ***Forward (Column Right*** or ***Column Half Right),*** and the other platoon leaders command ***Stand Fast.***

a. On the command of execution **MARCH,** the right platoon marches in the direction indicated. All other platoons follow (in sequence) in column, executing *Column Half Right* and *Column Half Left* on the commands of the platoon leaders.

b. To execute the movement when marching, the company commander commands ***Port,*** **ARMS** (unless at *Sling Arms),* and then commands ***Column of Platoons, Right Platoon, Double Time,*** **MARCH.** On the preparatory command, the right platoon leader gives the supplementary command ***Double Time,*** and the other platoon leader gives the supplementary command ***Continue to March.*** On the command of execution **MARCH,** the right platoon marches in *Double Time.* Other platoon leaders (in sequence) command ***Column Half Right, Double Time,*** **MARCH** and ***Column Half Left,*** **MARCH** to bring the succeeding platoons in column with the leading platoon.

c. The platoon leader and the platoon sergeant reposition themselves after the supplementary command but before the command of execution.

8-16. FORMING A COMPANY IN COLUMN WITH PLATOONS IN LINE AND RE-FORMING

This movement is normally executed with the commander at his post. For clarity purposes, the term "commander" is used. To form a company in column with platoons in line from a column formation at the *Halt,* the command is ***Column of platoons in line,*** **MARCH.** The platoon leader of the leading platoon gives the supplementary command ***Stand Fast.*** The second platoon leader gives the supplementary command ***Column Right.*** All other platoon leaders give the supplementary command ***Forward.***

a. On the command of execution **MARCH,** the leading platoon *Stands Fast,* and the second platoon executes a *Column Right,* marches 12 steps past the right file of the first platoon, and executes a *Column Left.* As they come on line with the base platoon, the platoon leader commands ***Mark Time,*** **MARCH.** After the platoon has marched in place for eight counts, the platoon leader commands ***Platoon,*** **HALT.** The succeeding platoons execute a *Column Right* At approximately the same location as the platoon to their front and execute a *Column Left* and then halt in the same manner as the second platoon. When the platoons have halted in position, the company commander commands ***Left,*** **FACE.** On that command, the platoon leaders and platoon sergeants face in marching and asume their posts.

b. When executed while marching, the movements are basically the same as from the *Halt* except that the commander gives the command of execution as the right foot strikes the marching surface. The leading platoon leader commands ***Mark Time.*** On the preparatory command, the second platoon leader commands ***Column Right,*** and the succeeding platoon leaders command ***Continue to March.*** On the command of execution **MARCH,** the leading platoon executes *Mark Time* and marches in place (approximately eight counts) until the platoon leader commands ***Platoon,*** **HALT**. The other platoons execute the movement in the same manner as from the *Halt.*

c. To re-form in column with platoons in column, the company commander commands ***Right, Face; Column of Platoons, Left Platoon,*** **MARCH.** On the command ***Right, Face,*** the platoon leaders and platoon sergeants face in marching and resume their posts in column. On the preparatory command ***Column of Platoons, Left Platoon,*** the left platoon leader commands ***Forward, (Column Left [Half Left]).*** All other platoon leaders command ***Column Half Left.*** On the command of execution **MARCH,** the left platoon executes the movement. The other platoon leaders give the appropriate commands so as to follow the lead platoon at correct distance.

NOTE: If necessary, the platoons following the second platoon automatically adjust the length of their step to ensure correct distance from the platoon to their front.

8-17. DISMISSING THE COMPANY

The company is dismissed while at *Attention.* It is usually dismissed by the first sergeant.

a. The first sergeant commands **TAKE CHARGE OF YOUR PLATOONS.** The platoon sergeants salute. The first sergeant returns all *Salutes* with one *Salute.* After *Salutes* are exchanged, the first sergeant and guidon bearer leave the formation. The platoon sergeants then dismiss their platoons as in platoon drill.

b. When the company is dismissed by the company commander, he commands **TAKE CHARGE OF YOUR PLATOONS.** The platoon leaders salute. The company commander returns all *Salutes* with one *Salute.* After the *Salutes* are exchanged, the company commander, guidon bearer, first sergeant, and executive officer leave the formation.

(1) The platoon leader(s) faces about and commands **PLATOON SERGEANT**. The platoon sergeant faces to the right in marching and marches (inclines) around the squad leader(s), halts three steps in front of and centered on the platoon leader, and faces to the right. Each platoon leader then directs **TAKE CHARGE OF THE PLATOON.** *Salutes* are exchanged. The platoon leaders leave the formation.

(2) The platoon sergeants step forward three steps, face about, and dismiss the platoons as in platoon drill.

Section III. INSPECTIONS

This section discusses the procedures used to conduct inspections for a company drill.

8-18. BASIC INFORMATION

The following basic information applies to conducting inspections for a company drill.

a. The company has one prescribed formation for inspecting personnel and equipment in ranks—company in line with platoons in line. When inspecting crew-served weapons and vehicles, the personnel are normally positioned to the rear of the formation with the operator (gunner) standing by his vehicle (weapon).

NOTES: 1. If the commander is not scheduled to receive or inspect the company, the first sergeant and platoon sergeants remain at their posts and execute company drill from their posts. For continuity purposes, the term "first sergeant" may also be used to denote the commander, just as "platoon sergeant" may be used to denote the platoon leader when they are executing drill from their posts. If the commander is inspecting, he and the platoon leaders execute the same as described below.

2. Manual of arms movements for soldiers armed with the M249, shotgun, or pistol are outlined in Appendix E.

b. The first sergeant may conduct an in-quarters (barracks) inspection to include personal appearance, individual weapons, field equipment, displays, maintenance, and sanitary conditions.

c. When field equipment is to be inspected, it should be displayed as shown in Figure 8-5, page 8-18, (as a guide). Additional equipment not shown, or different models of the equipment, should be arranged in a uniform manner established by the local commander.

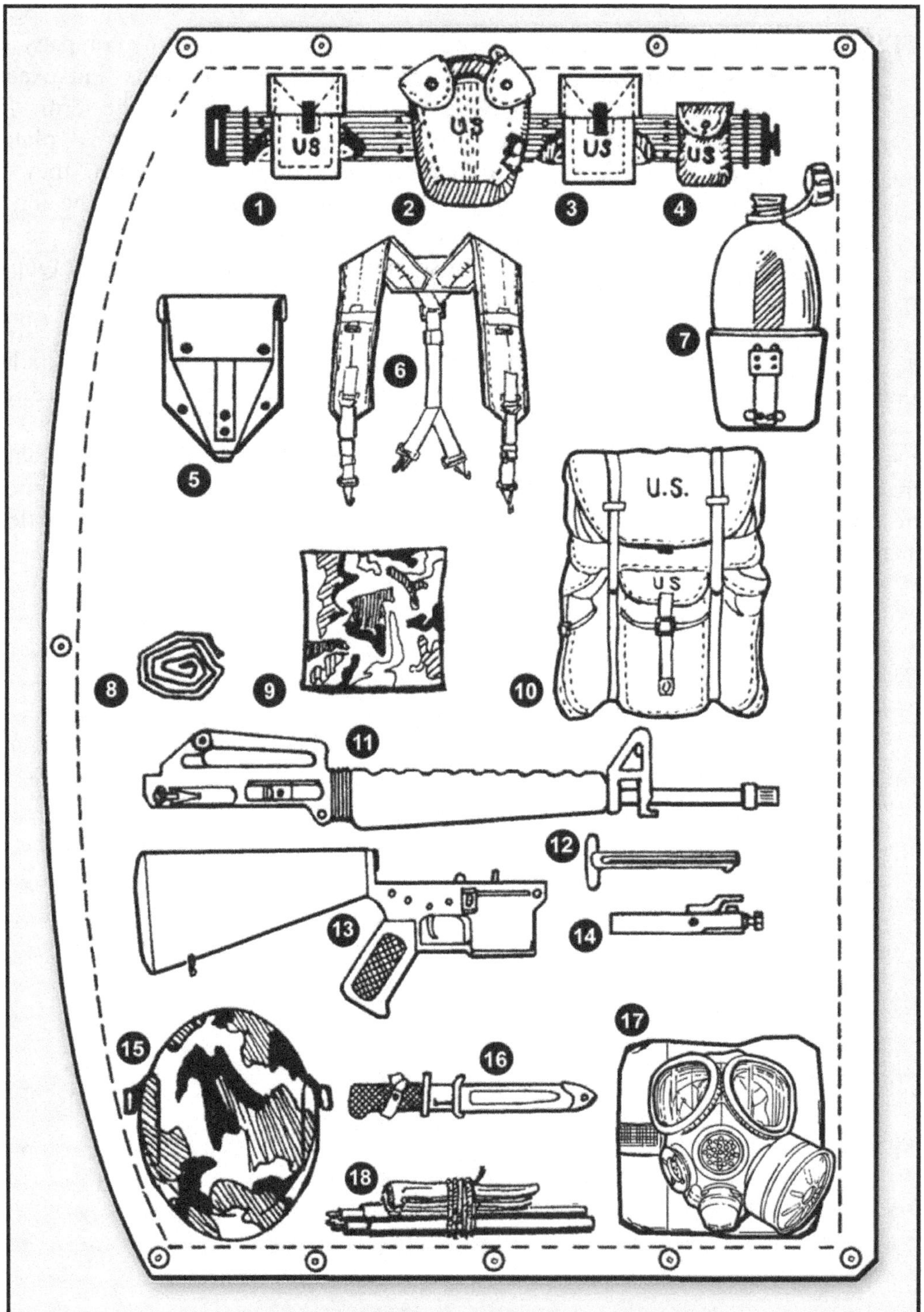

Figure 8-5. Equipment display (guide).

1. PISTOL BELT	10. RUCK SACK
2. AMMUNITION POUCH	11. UPPER RECEIVER
3. CANTEEN COVER	12. CHARGING HANDLE
4. AMMUNITION POUCH	13. LOWER RECEIVER
5. ENTRENCHING TOOL AND CARRIER	14. BOLT CARRIER GROUP
6. FIELD SUSPENDERS	15. HELMET
7. CANTEEN	16. BAYONET AND SCABBARD
8. SLING	17. PROTECTIVE MASK
9. PONCHO	18. TENT STAKES, POLES, AND ROPE

Figure 8-5. Equipment display (guide) (continued).

8-19. IN-RANKS INSPECTION (LINE FORMATIONS)

Use the following procedures to conduct an in-ranks inspection (line formation).

a. With the company in line formation, the first sergeant directs **PREPARE FOR INSPECTION.** On that directive, the platoon sergeants face about, open ranks, and align the company as in paragraphs 8-5 and 8-6. When all platoon sergeants are facing the front, the first sergeant commands **AT EASE.**

NOTE: During the inspection, the guidon bearer, officers, and noncommissioned officers not in ranks assume the position of attention as the inspecting officer approaches their positions and resume the at ease position after they have been inspected (they do not execute inspection arms). The company commander may direct the first sergeant and executive officer to accompany him during the inspection. When so directed, the executive officer and first sergeant take a position at normal interval (close interval if the company is at close interval) to the left of the company commander.

b. The first sergeant begins the inspection by stepping forward and inspecting the guidon bearer. He then faces to the half left in marching and proceeds to the right of the line. As he approaches the right flank platoon, the platoon sergeant commands (over his right shoulder) ***Platoon,*** **ATTENTION.** When the first sergeant has halted directly in front of him, he reports, ***"First Sergeant, the platoon is prepared for inspection."*** (If the commander is inspecting, *Salutes* are exchanged.) The first sergeant inspects the platoon sergeant. The first sergeant then directs the platoon sergeant to lead him through the inspection, faces to the half left in marching, and halts directly in front of the squad leader of the first squad. As the first sergeant faces to the half left in marching, the platoon sergeant faces to the right in marching, inclines, and halts directly in front of the second man in the first squad and on line with and at *Normal Interval (Close Interval)* to the right of the first sergeant. The other platoon sergeants execute the same actions as the first sergeant approaches their platoons.

NOTE: As soon as the platoon sergeant and first sergeant have halted in front of the first two men, the platoon sergeant commands ***Second, Third, and Fourth Squads,*** **AT EASE**. When moving from man to man during the inspection, the first sergeant and platoon sergeant (executive officer and first sergeant if the commander is inspecting) simultaneously face to the right in marching, take two short steps (one step if the company is at close interval), halt, and face to the left. After the first sergeant has inspected the last man in the front rank, the platoon sergeant hesitates momentarily and allows the first sergeant to precede him as he inspects the front rank from the rear. As the first sergeant begins to inspect the rear of each rank, the platoon sergeant commands the next squad to *Attention*. As they begin to inspect the next rank, the platoon sergeant commands the last rank inspected to stand *At Ease*. Normally, when inspecting the rear of each rank, the first sergeant conducts a walking inspection. As the first sergeant inspects the rear of the last man in each rank, he turns and halts directly in front of the squad leader of the next rank. The platoon sergeant turns and halts directly in front of the second man. If the commander is inspecting, the executive officer and first sergeant march past (behind) the company commander and assume their positions to his left.

c. As the first sergeant halts and is directly in front of and facing the individual being inspected, the individual executes *Inspection Arms*. If the first sergeant wants to inspect the individual's weapon, he inspects the weapon first, then the individual's uniform and appearance. As soon as the first sergeant grasps the weapon, the individual releases the weapon and resumes the *Position of Attention*. When the first sergeant has finished inspecting the weapon, he returns it in the same manner as it was received. The individual receives the weapon by grasping the center of the handguard just forward of the slip ring with the left hand and the small of the stock with the right hand. He then executes *Ready, Port Arms; Order Arms.*

NOTE: In the event the first sergeant does not want to inspect the weapon, the individual remains at *Inspection Arms* until the first sergeant has halted in front of and is facing the next man. The individual then executes *Ready, Port, Arms* and *Order Arms*. As the first sergeant moves to the rear of the rank being inspected, each man, as the first sergeant approaches, unfastens the snap of his scabbard, grasps the point of the scabbard with the left hand, and moves the point slightly forward. The scabbard is held in position until the bayonet has been replaced or, if the bayonet is not inspected, until the first sergeant has moved to the next man. At that time, the securing straps are fastened and the *Position of Attention* is resumed.

d. The correct method of manipulating a rifle for inspection is as follows:

(1) When receiving the rifle from the inspected individual, reach forward and grasp the rifle at the upper portion of the handguard with the right hand. Lower the rifle diagonally to the left, twist the rifle slightly, insert the tip of the little finger of the left hand into the ejection port (do not twist the rifle but insert the thumb into the receiver of the M14-series rifle), and look into the barrel. The finger reflects sufficient light for the inspection of the barrel.

(2) Grasp the small of the stock with the left hand and raise the rifle to a horizontal position (sights up) with the muzzle to the right. The rifle is centered on the body with

the forearms horizontal and the palms up. Move the rifle horizontally to the left and inspect the upper portion of the rifle beginning with the flash suppressor. Move the rifle slowly to the right inspecting the butt. Return the rifle to the center of the body.

(3) Keeping the rifle horizontal, rotate it 180 degrees (sights toward the body) so that the sights point toward the marching surface. Move the rifle horizontally to the left and inspect the muzzle. Move the rifle slowly to the right and inspect the butt. When moving the rifle across the body, twist the rifle as necessary and inspect the movable parts on the sides of the receiver. Having completed the inspection of the lower parts of the rifle, return it to the center of the body.

(4) Keeping the rifle horizontal, rotate the rifle 180 degrees (sights away from the body) so that the sights are up. Release the left hand and return the rifle in the same manner as received.

(5) When inspecting a weapon other than a rifle, the inspector receives the weapon with his right hand and inspects it in the most convenient manner. Having completed the inspection, the weapon is returned with the right hand in the same manner as received.

e. When the first sergeant has completed the inspection of the platoon, the platoon sergeant commands the platoon to *Attention* and overtakes the first sergeant en route back to the front of the platoon. The platoon sergeant halts at his post facing to the front (*Salutes* are exchanged if the commander is inspecting). The first sergeant (or commander with or without the inspecting party) faces to the right in marching and moves to the next platoon. As soon as the first sergeant has cleared the first platoon, the platoon sergeant faces about, commands ***Close Ranks,*** **MARCH** and **AT EASE,** and then faces about and executes *At Ease.* When the first sergeant has completed the inspection of the last platoon (exchanged *Salutes* with the platoon sergeant), he returns to his post at the center of the company and commands the company to *Attention.* He then dismisses the company, as previously described.

f. When the company is inspected by an officer of a higher command, on the approach of the inspecting officer, the company commander commands ***Company,*** **ATTENTION.** He then faces about, salutes, and reports, ***"Sir, the company is prepared for inspection."*** The inspection is conducted as previously described except that the company commander takes a position immediately to the left of the inspecting officer. The inspection is terminated by the company commander in the same manner prescribed for a platoon sergeant.

8-20. IN-QUARTERS INSPECTION (STAND BY)

When the first sergeant wants to conduct an in-quarters inspection, the members are positioned on line with their equipment (or as near it as possible).

a. The platoon sergeant positions himself in the path of the inspecting officer at a point near his area of responsibility. At the approach of the inspecting officer, the platoon sergeant commands ***Platoon,*** **ATTENTION,** salutes, and reports, ***"Sir, the platoon is prepared for inspection."***

NOTE: When the situation dictates, the platoon sergeant may report to the inspecting officer outside of the quarters. A designated individual commands the members to *Attention* as the inspector enters the quarters (bay).

b. The platoon sergeant then guides the inspector along a route dictated by the physical arrangement of the personnel and equipment. Upon entering the area, the highest ranking soldier present commands **AT EASE**. As the inspector approaches each individual or his equipment, the individual automatically assumes the *Position of Attention*. When the inspector has moved to the next man, the last man inspected resumes the position of *At Ease*. As the inspector begins to exit the area after completing the inspection, **CARRY ON** is commanded.

NOTE: When inspecting soldiers' rooms, *Attention* is not commanded. The individuals automatically assume the *Position of Attention* as the inspector enters the room. The inspector then commands **AT EASE**. As the inspector approaches each soldier, the soldier assumes the *Position of Attention* and resumes *At Ease* after he has been inspected.

CHAPTER 9
BATTALION AND BRIGADE DRILL

"The exterior splendor, the regularity of movements, the adroitness and at the same time firmness of the mass—all this gives the individual soldier the safe and calming conviction that nothing can withstand his particular regiment or battalion."

Colmar von der Goltz, 1843-1916 Rossback und Jena

For the most part, battalion and brigade drill merely provides the procedures for executing company or battalion drill in conjunction with other companies or battalions in the same formation.

For drill purposes, a battalion or brigade consists of a headquarters (staff), Colors, and two or more companies or battalions. The right flank unit serves as base when in a line formation; the leading element is the base when in column.

NOTE: Executive officers, first sergeants, and platoon sergeants form the same as in company drill for all formations unless otherwise specified.

9-1. BATTALION FORMATIONS

The battalion has two basic formations—a line and a column. Separate elements may be arranged in several variations within either formation: the battalion may be formed in line with the companies in line with platoons in line, or battalion in line with companies in column with platoons in line (Figure 9-1, page 9-2). From those formations, the battalion may be positioned in a battalion in column with companies in column, or companies in mass, or companies in column with platoons in line (Figure 9-2, page 9-3).

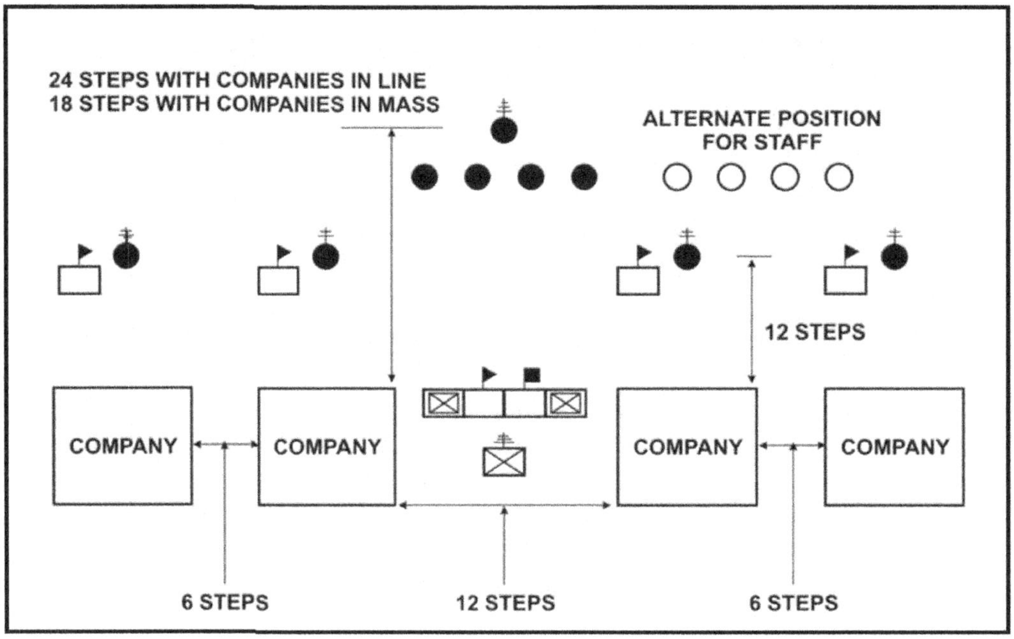

Figure 9-1. Battalion in line with companies in line or mass.

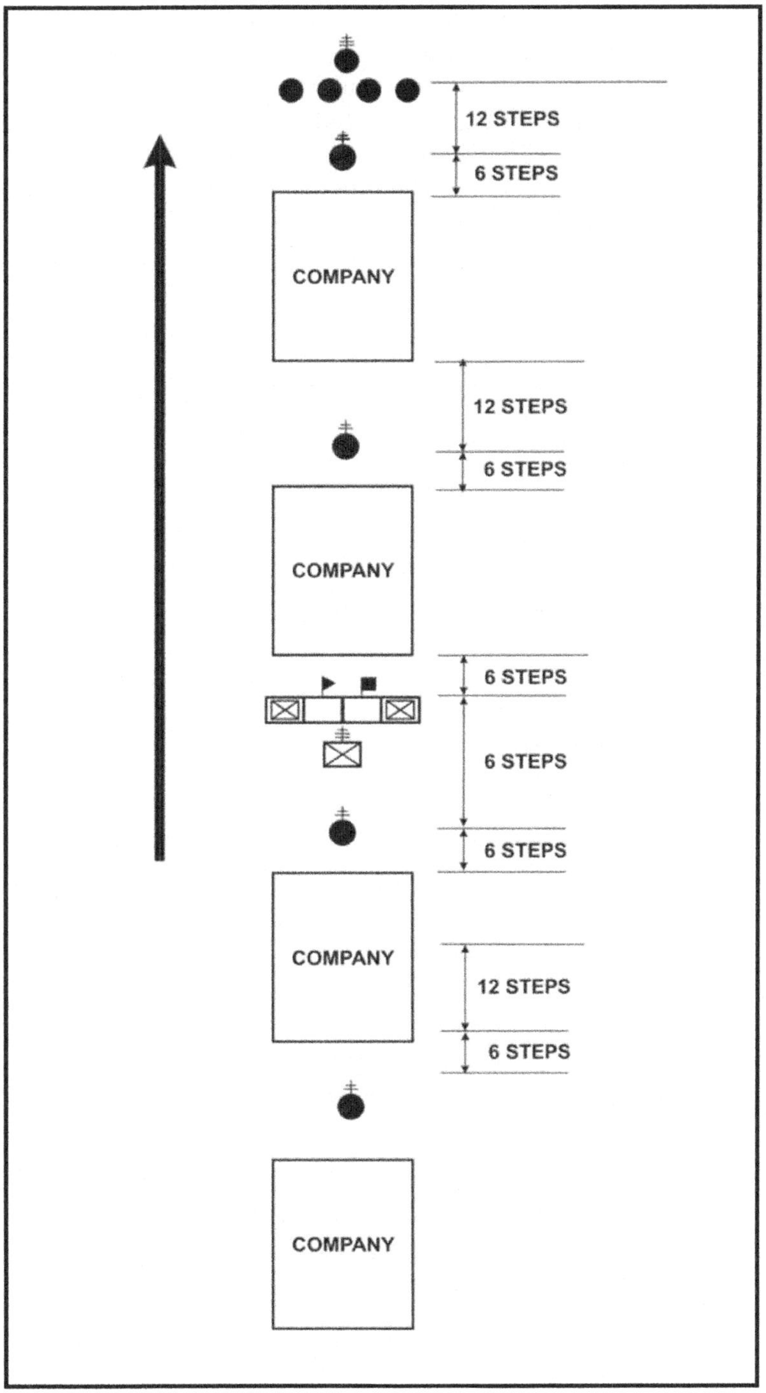

Figure 9-2. Battalion in column with companies in column or mass.

a. **Forming the Battalion.** To facilitate the forming of a larger unit, the commander normally alerts the component units as to the desired formation, time, place, route, uniform, and the sequence in which the units will form. Unless the right guide's position at the formation site has been previously marked, the guides report to the site and receive instructions before the arrival of the troops.

(1) The commissioned staff forms in one rank, at normal interval and centered on the commander. Staff members are normally arranged in their numerical order from right to left. When enlisted staff personnel form as part of the officer staff, they form two steps to the rear of their respective staff officer. The command sergeant major forms one step to the rear and centered on the Colors.

(2) When the battalion participates as a separate element of a larger formation, or when space is limited, it is normally formed in a mass formation in the same manner as forming a company mass (Figure 9-3).

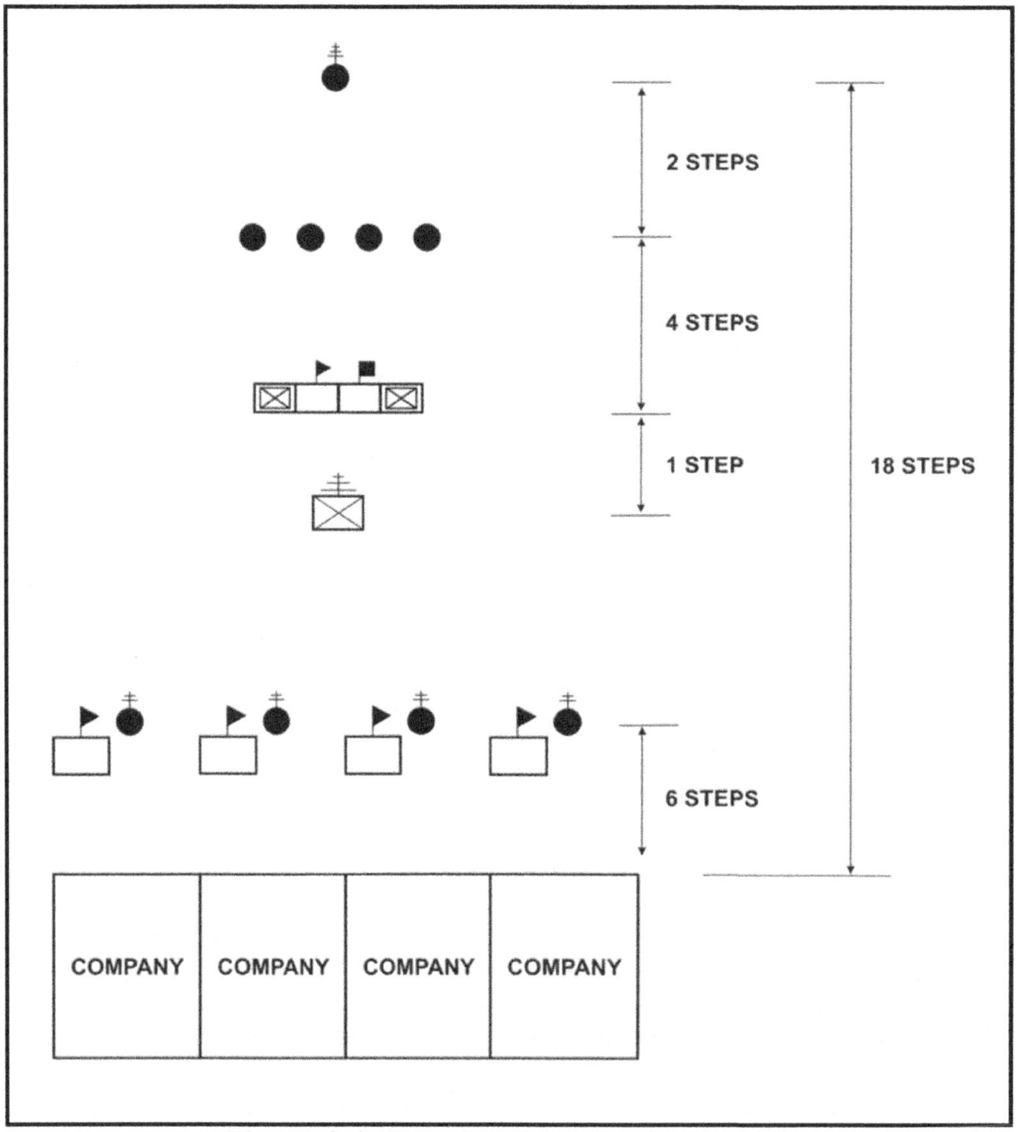

Figure 9-3. Battalion in mass formation.

(3) The adjutant is responsible for the formation of troops. He takes a position at the right flank of the line on which the units are to form and faces down the line. He remains facing down the line until all units are formed and then marches to his post midway between the line of troops and the proposed position of the commander. As the commander and his staff approach, the adjutant commands (directs) the units to **ATTENTION,** faces the commander, salutes, and reports, ***"Sir, the battalion is formed."***

NOTE: If guides are used, the adjutant takes his post as soon as the guides have been aligned.

(4) The commander returns the *Salute* and commands **POST.** The adjutant marches forward, passes to the commander's right, and takes his post as the right flank staff officer. The commander then commands or directs such actions as he desires.

(5) Normally, the Colors are positioned at the center of the battalion when formed in line or column, and four steps to the rear of the staff when formed in mass.

b. **Dismissing the Battalion.** To dismiss the battalion, the commander commands **TAKE CHARGE OF YOUR UNITS.** The company commanders and battalion commander exchange *Salutes.* The battalion commander returns all *Salutes* with one *Salute.* He then dismisses his staff.

NOTE: In a battalion mass formation, the platoon leaders form as the first rank of troops. The right flank platoon leader serves as the guide.

c. **Inspecting the Battalion.** When the battalion is to be inspected, the adjutant forms the battalion in line with companies in line and platoons in line.

(1) The adjutant forms the troops, reports, and takes his position with the staff (as previously described). When the adjutant has joined the staff, the battalion commander directs **PREPARE FOR INSPECTION.** Unit commanders face about and give the same directive. The platoon leaders prepare their platoons for inspection in the same manner as in company drill. When all of the platoons in each company have completed opening ranks and have dressed right, the company commander faces the battalion commander. When all company commanders are again facing him, the battalion commander commands **AT EASE**.

NOTE: Because of the time involved in inspecting a battalion, the commander normally directs his staff to inspect the companies while he inspects the Colors and makes a general inspection of the battalion. The Color guards do not execute *Inspection Arms*.

(2) As the battalion commander (or his designated staff officer) approaches the company, the company commander faces about, salutes, and reports, ***"Sir, _____ Company is prepared for inspection."*** The inspecting officer then proceeds to the first platoon to be inspected. The company commander takes a position to the left of the inspecting officer. The platoons are inspected in the same manner as explained in company drill. In the event a platoon has already been inspected by a staff officer, or is

waiting to be inspected by a staff officer, and the battalion commander approaches, the battalion commander is invited to inspect the platoon.

(3) When the inspection has been completed and the battalion commander is en route back to his post, the company commanders bring their units to **ATTENTION** and **CLOSE RANKS** and **AT EASE.** When the battalion commander has returned to the front of the formation and all staff officers have completed their inspection, the battalion commander commands ***Battalion,*** **ATTENTION** and dismisses the battalion (as previously described).

(4) If the battalion is being inspected by a higher commander and his staff, the battalion commander reports to the inspecting officer in the same manner as the company commanders. The battalion commander's staff takes a position to the left of the higher inspecting staff and escorts them to the unit. The battalion commander escorts the higher commander.

9-2. BRIGADE FORMATIONS

The brigade, like the battalion, has two basic formations—a line and a column. The battalions may be arranged in several variations within either formation: the brigade may be formed with battalions in mass, or battalions in line with companies in mass. From those formations, the brigade may be positioned in column with battalions in mass, companies in mass, or battalions in column with companies in column. The formation selected is normally dictated by the space available and the desires of the commander (Figure 9-4).

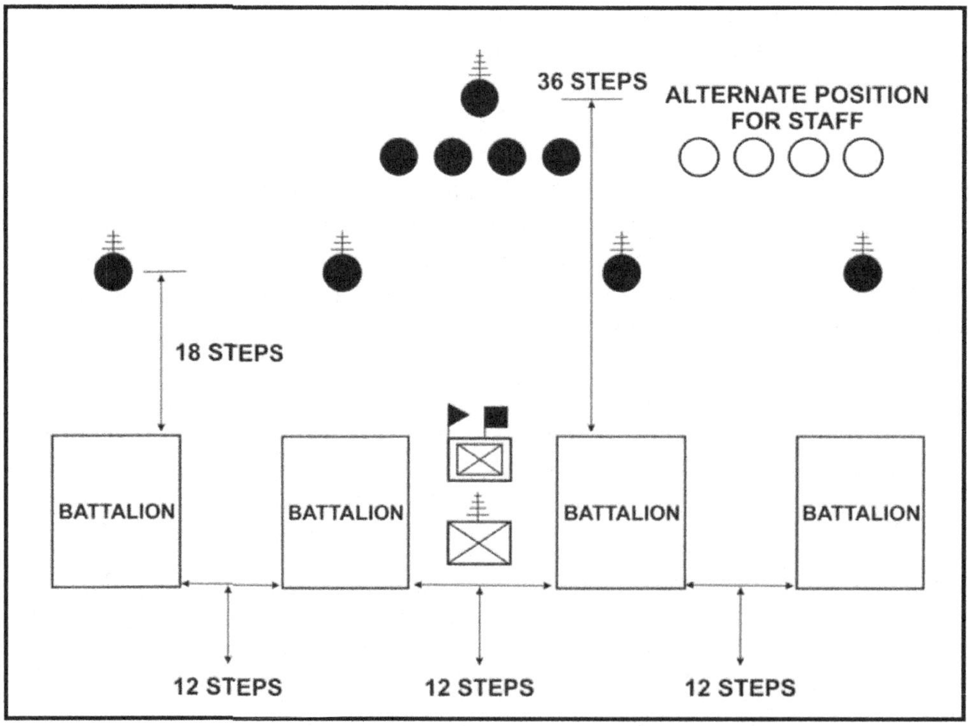

Figure 9-4. Brigade in line with battalions in mass.

a. The brigade is formed and dismissed in the same manner as the battalion.

b. Normally, the brigade Color is positioned at the center of the brigade. Subordinate Color-bearing organizations position their Colors to the center of their formation when in a line or column formation, or four steps to the rear of their staff when in a mass formation (Figure 9-5).

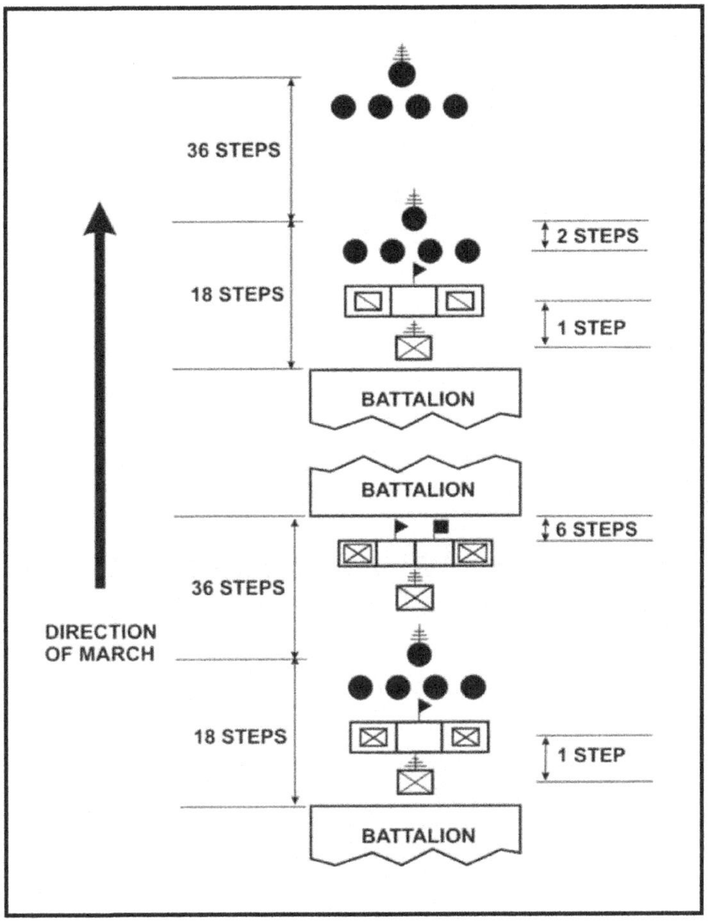

Figure 9-5. Brigade in column.

This page intentionally left blank.

PART TWO
CEREMONIES

"A ceremonial parade, impeccably performed, can never fail to be a source of inspiration to those who watch it or take part in it. It is the noblest and proudest form of drill. Based on the 'blunt whetstone' of drill instruction to recruits it was for many hundreds of years the foundation of battle discipline in all Armies . . .

"Today, once the elements of discipline have been instilled through drill on the parade square, it develops, naturally, into various forms of crew drill, gun drill, and battle drill but the aim of discipline remains unchanged. This aim is the conquest of fear. Drill helps to achieve this because when it is carried out men tend to lose their individuality and are unified into a group under obedience to orders.

"If men are to give their best in war they must be united. Discipline seeks through drill to instill into all ranks this sense of unity, by requiring them to obey orders as one man. A Ceremonial parade, moreover, provides an occasion for men to express pride in their performance, pride in the Regiment or Corps and pride in the profession of Arms."

General Sir Harold Alexander, Ceremonial for the
Army Ministry for Defense, Army Department,
November 1968

NOTE: See Appendix J for a checklist for conducting a successful ceremony.

CHAPTER 10
REVIEWS

This chapter discusses the general history of the review and the different types of reviews used in the Army today.

10-1. HISTORY

Reviews can be traced to the Middle Ages when rulers, as a way of showing their strength, were likely to have military ceremonies.

a. In the American Army, reviews were originally outlined in Baron Friedrich von Steuben's Blue Book and practiced by revolutionary soldiers. A review consisted of four stages: a formation of troops, presentation and honors, inspection, and a march in review. Today's reviews have incorporated three additional stages: honors to the nation, remarks, and a conclusion. The presence of the band represents the significant role that the drum, fife, and other musical instruments have played throughout military history for signaling in camp or on the battlefield. The presence of the Colors at the center of the formation represents their presence at the forefront of the unit during the heat of battle.

b. At the beginning of the 17th century, armies throughout the world were adopting the regimental system. Regiments were assigned a specific color or number for ease of identification and position on the battlefield. In battle, the color (or flag) party marched at the front and center of its unit as a point for the unit to dress on. By leading the unit in battle, the Colors became prime targets, as victories in those days were expressed in terms of the number of enemy Colors captured; consequently, the Color party bore the brunt of the battle and suffered heavy casualties. Historically, in the American Army, the Colors have been placed in the center of the formation and have been considered to be a part of the company on the immediate right of the Colors. This company is still called the Color unit.

(1) In infantry regulations of 1812, the line up of companies from right to left was determined by the seniority of the company commander, with the senior on the right. The manual for U.S. infantry tactics in 1862 placed the senior commander on the right and the second senior on the left with the other eight companies in between. After they were lined up, the companies were then numbered first through tenth from right to left. In the infantry drill regulations of 1892, companies continued to be aligned by seniority. Between 1910 and 1970, companies were designated by letters of the alphabet and lined up from right to left: headquarters company, rifle companies in alphabetical order, then any combat support companies. The Colors, however, remained posted in the center of the formation.

(2) Musical elements (usually drums and fifes, and sometimes a band) were used during actual battle and marched 12 to 15 paces to the rear of the formation. By 1892, the infantry drill regulations posted bands and field music on the right of the formation or at the head of the formation when it moved in column.

c. Honors appear to have originated as musical fanfare and later a gun salute was added. The combination of the two now constitute military honors.

(1) The custom of announcing the arrival of royalty and heads of state with a fanfare of trumpets or drum rolls had its origin in England. It is ironic that drum *ruffles* and the two-note bugle call *flourishes,* used in this manner today, have all but disappeared in the British Commonwealth while the tradition is still carried on in the United States.

(2) The colonists brought many of the military customs of their mother country with them when they came to America. At the time of the French-Indian Wars, soldiers were instructed to "rest (present arms) and beat two ruffles" for the Royal Governor of Virginia. In 1776, officers of the Continental Army were advised that "the adjutant is to order a drum and fife to give two ruffles to a major general and one to a brigadier."

(3) When Baron von Steuben wrote his manual of discipline, he stated that there were to be "certain ruffles" for generals. It is interesting to note that General George Washington was not accorded ruffles. Instead, his arrival was announced by a march—more than likely "Washington's March"—that is still played today. In time, this custom was extended to include governors of sovereign states and visiting foreign dignitaries.

(4) Ruffles were not considered elegant and were used exclusively by the military. In some cases they were even accorded to the officer of the day.

(5) It was some time before trumpets were added to the drum rolls; it is believed they were so used during the War of 1812. Honors for dignitaries were modified to include a flourish of trumpets called the "Three Cheers." It is likely that this particular salute evolved into the ruffles and flourishes as they are played today.

(6) Navy records indicate that the regulations of 1876 stipulated there would be a ruffle of drums to announce the President's arrival but made no mention of the trumpet flourishes. A later regulation (1893) stated that the chief executive would receive four ruffles and flourishes.

(7) Gun salutes may be traced back to a period when it took a long time to reload guns. By firing all of its guns at once, the battery or fort was left defenseless. The number of guns fired as a salute to honor a U.S. President was not established until 1841. Before that, presidential honors included one gun for every state. It reached at least 24 guns before it was decided that any more guns would be unrealistic and the number of 21was adopted in 1841. In 1875, the United States adopted 21 guns for the international salute and that number is currently used as honors for all dignitaries of rank equivalent to the President.

10-2. TYPES OF REVIEWS

A review is a military ceremony used to—

- Honor a visiting, high-ranking commander, official, or dignitary, and or permit them to observe the state of training of a command.
- Present decorations and awards.
- Honor or recognize unit or individual achievements.
- Commemorate events.

a. Normally, a review is conducted with a battalion or larger troop unit; however, a composite or representative element consisting of two or more platoons may serve the same purpose. A review consists of the following steps in sequence:

- Formation of troops.
- Presentation of command and honors (although primarily a characteristic of a parade, "sound off" may be included here to enhance the ceremony).
- Inspection. (The inspection stage may be omitted for decorations, awards, or individual recognition ceremonies.)
- Honors to the nation.
- Remarks.
- March in review.
- Conclusion.

NOTE: If retreat is conducted as part of the review, it occurs after presentation of the command and honors and before the inspection. Honors to the nation is omitted.

b. Other ceremonial activities, or combinations thereof, that may be incorporated within the framework of a review are:

- Review with decorations, awards, and individual retirement.
- Review with change of command, activation, or inactivation of units.
- Review with retreat.
- Review with retreat and retirement, decorations, and awards.
- Review with retreat and change of command, activation, or inactivation of units.

NOTE: Review with change of command should not be combined with an awards ceremony. Awarding a decoration to an outgoing command is accomplished separately from, and preferably before, a review at which the responsibility of command is transferred.

c. In order that he may review his own command or accompany a visiting reviewing officer, a commander normally designates an officer of his command as commander of troops. The commander of troops is responsible for the preparation of the troops for the review.

d. The ready line and final line on which the units are to form and the route of march are marked or designated before beginning the ceremony (Figure 10-1, page 10-6).

(1) The primary function of the ready line is to enable the unit commander to arrange his unit into the prescribed formation before movement to the final line. The ready line is to the rear of the final line. Flags or appropriate markings are used to designate the post of the reviewing officer and the points where ***Eyes,*** **RIGHT** and ***Ready,*** **FRONT** are to be commanded (6 steps to the reviewing officer's right and left and from 6 to 20 steps in front of the reviewing officer).

(2) Commanders should determine the length of their formation (steps) so that they know how far beyond the ready front marker (6 steps beyond the reviewing officer) they command ***Ready,*** **FRONT**. This ensures that the last members of their unit are six steps beyond the reviewing officer when the command is given.

(3) The command ***Eyes*** is given as the right foot strikes the marching surface, and the command of execution **RIGHT** is given the next time the right foot strikes the marching surface. The commands ***Ready*** and **FRONT** are given as the left foot strikes the marching surface.

NOTE: These commands are given when the commander reaches the guide marker. The guide marker is placed so that when the commander reaches it, the last element of his unit is abreast of the ready front marker.

e. Any of the formations described for the battalion or brigade may be used; however, the two recommended formations for conducting reviews are: battalion in line with companies in mass (Figure 10-2, page 10-7), or brigade in line with battalions in mass (Figure 10-3, page 10-7).

NOTE: When desired or more appropriate, commands may be substituted for directives. When desired or appropriate, the formation may perform mounted using the mounted drill outlined in Appendix K.

f. The formation selected is determined by space available and other desires of the commander. Commanders may alter the formation or prescribed distances to meet local situations. Each unit should be sized uniformly with the tallest men in front and on the right. Commanders should not cause the leaders to change positions because of their size. Officers and key noncommissioned officers, including squad leaders and others equivalent by virtue of their rank and time in service, should participate in their deserved positions of dignity.

g. If possible, the reviewing officer should arrive at his post (Figure 10-4, page 10-8) after the staff of the commander of troops has reversed and the commander of troops has faced the reviewing stand. Upon arrival, the reviewing officer should immediately take his post and refrain from greeting distinguished spectators until after the review has terminated.

NOTE: In battalion mass formation, the platoon leaders form as the first rank of troops. The right flank platoon leader serves as the guide.

h. The reviewing officer, host or host commander, and distinguished persons invited to attend the review (but not themselves receiving the review), take positions facing the troops (Figure 10-4, page 10-8).

i. When a ceremony is conducted for an individual junior in rank to the host or host commander, the junior takes a position to the left of the host or host commander unless the commander yields the post of honor to him. In all cases, the honor position is the position on the right and is the reviewing officer's post. The host or host commander gives the appropriate directive, **PRESENT THE COMMAND** and **PASS IN REVIEW**, during the conduct of the review.

j. When a civilian or foreign dignitary receives the review, he takes the honor position to the right of the host. The host or host commander returns all *Salutes* when, in his judgment, it would be more appropriate. As a courtesy, the host or host commander should cue the dignitary and inconspicuously explain the procedures during the conduct of the ceremony. The position in which the flags are displayed at the rear of the reviewing party corresponds to the position taken by the individual in the front rank of the reviewing party.

k. When a ceremony is conducted honoring a foreign official or dignitary entitled to honors, equivalent honors *(Salutes)* are rendered in the manner prescribed in AR 600-25.

NOTE: If the honors of the dignitary include his or her nation's anthem, the honors are played in the following order: four ruffles and flourishes, the foreign anthem, an approximate 3-second drum roll, and the U.S. National Anthem.

l. When the ceremony includes decorations, awards, or retirements, the host or host commander should be senior in rank or position to the highest ranking individual being honored.

m. An officer from the local staff is designated to escort and brief members of the reviewing party and to show them their positions.

n. Since all situations or eventualities relative to various services or units (military academies, special honor guard, active Army units, reserve components, Reserve Officer Training Corps, and training centers) cannot be foreseen, commanders may make minor changes to stated procedures. However, with a view toward preserving Army tradition and maintaining consistency, they should not alter or modify the sequence of events other than stated.

NOTE: Ceremonies may be conducted with sabers and swords. See Appendix F for the manual of arms for these weapons.

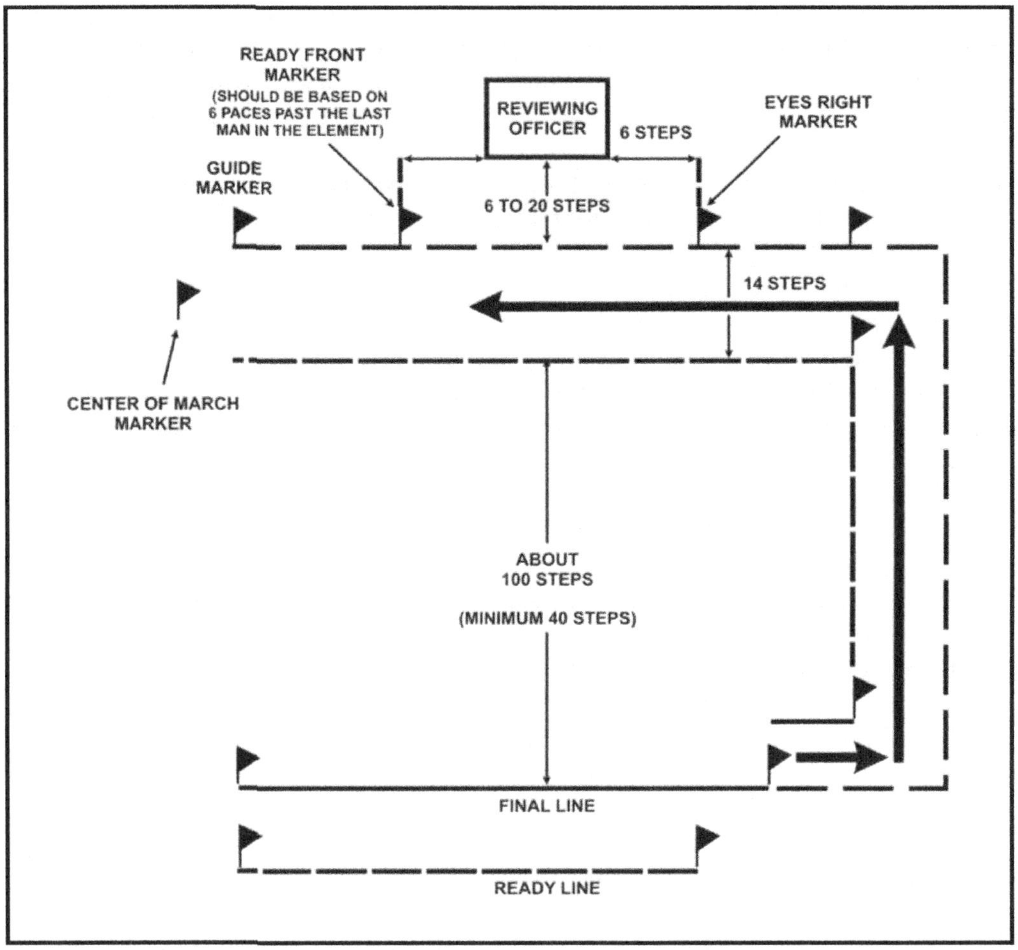

Figure 10-1. Preparation for review.

NOTE: The ready front marker is always six steps to the right of the reviewing officer. The commander uses this point to estimate the placement of the guide marker. The guide marker is used as a visual reference to the commander so that when he gives the command of execution **FRONT** of ***Ready,*** **FRONT,** the last members of the unit are at the ready front marker.

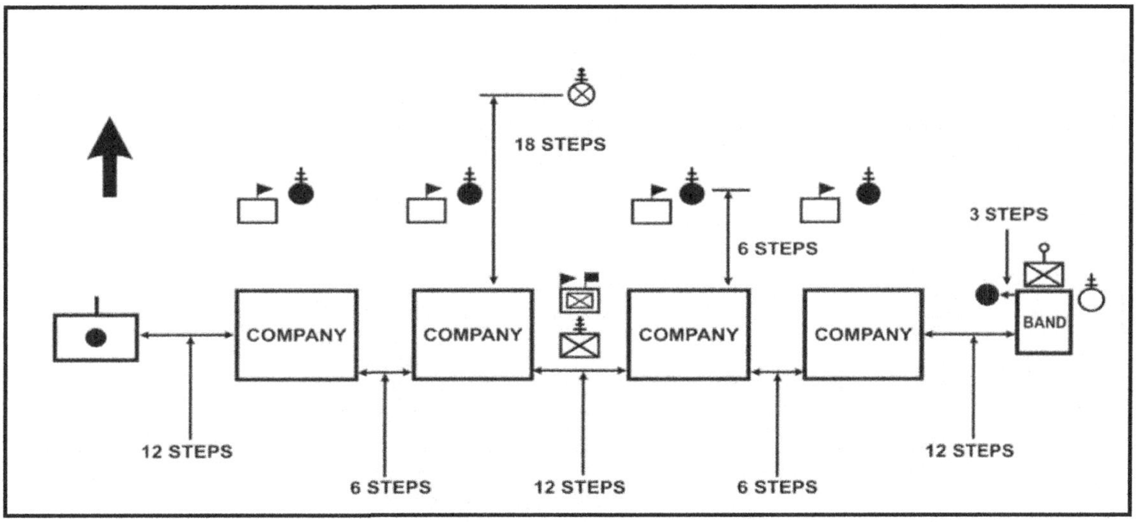

Figure 10-2. Battalion in line with companies in mass.

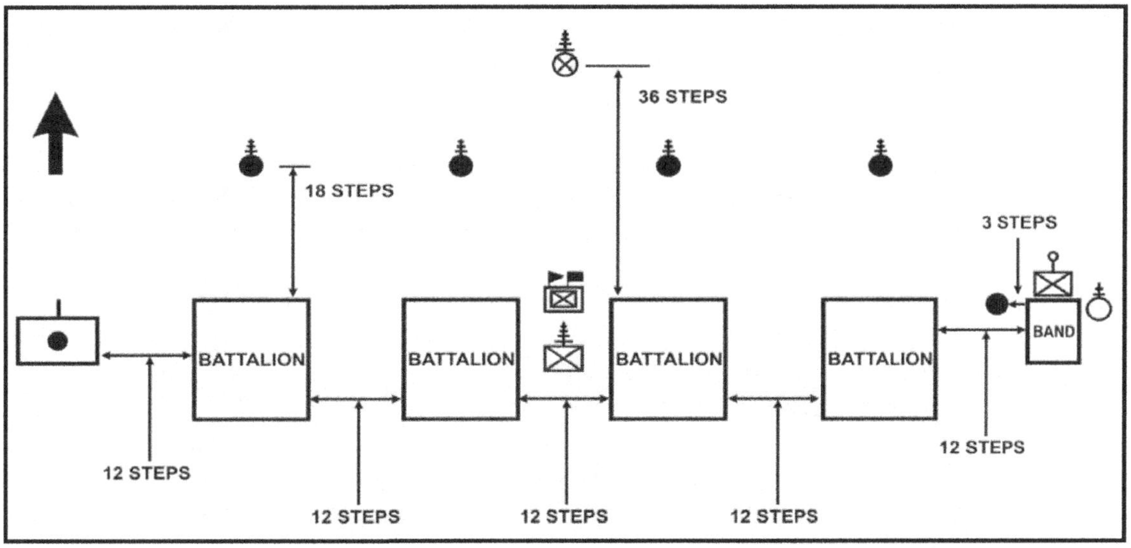

***Figure 10-3. Brigade in line with battalions in mass.**

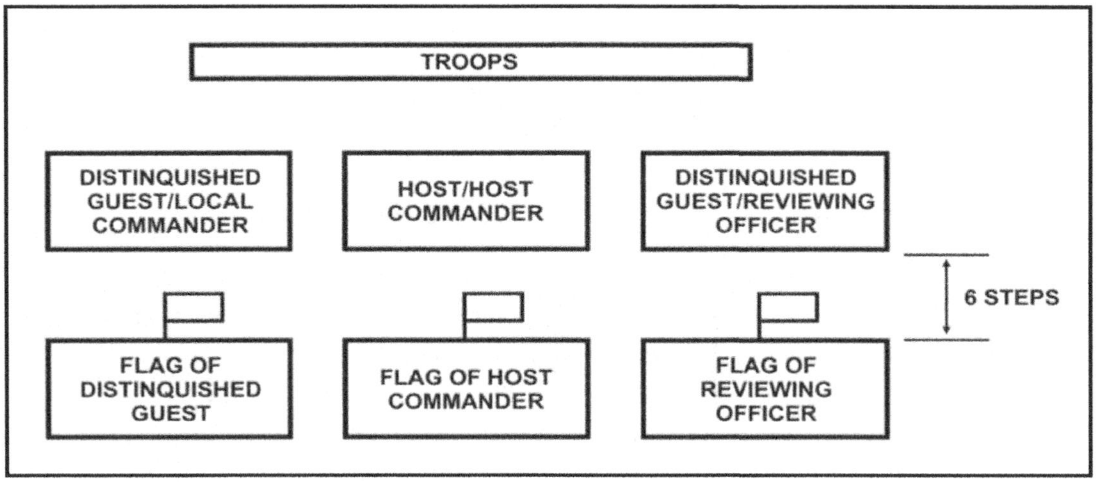

Figure 10-4. Reviewing party.

10-3. SEQUENCE OF EVENTS AND INDIVIDUAL ACTIONS

The sequence of events in conducting a review and the individual actions that take place are described herein.

a. **Formation of Troops.** Units are moved to their positions on the ready line in the most convenient manner. For a large review, the commander of troops prescribes the routes, sequence, and time of arrival on the ready line.

NOTE: If the ready line is not used, the units form on the final line. They are aligned and given the command ***Parade,*** **REST**. To alert the units that the ceremony is to begin, the adjutant (from his position near the band) directs **SOUND ATTENTION**. The band immediately sounds *Attention.* Unit commanders immediately face about and command their units (in sequence from right to left) to **ATTENTION** and then face about. When the last unit is at *Attention*, the adjutant directs **SOUND ADJUTANT'S CALL**. The band immediately sounds *Adjutant's Call.* The adjutant then moves to his post midway between the line of troops and the designated post of the commander of troops and faces the commander of troops. When the adjutant has halted at his post, the commander of troops marches his staff to their post midway between the line of troops and the reviewing officer's post. The ceremony then continues as prescribed in this paragraph.

(1) In motorized and mechanized units, occupants of vehicles form dismounted in a formation corresponding to that of the other units. Their vehicles remain to the rear of the formation.

(2) The band takes its position on the final line 12 steps to the right of the right flank unit's marker. The adjutant's initial post is three steps to the left of the band. After verification that all units are positioned on the ready line, the adjutant faces down the final line and over his right shoulder directs the band to **SOUND ATTENTION**.

(3) The band sounds *Attention,* pauses, and waits for the directive to **SOUND ADJUTANT'S CALL**.

(4) Upon hearing *Attention,* unit commanders *Face About,* bring their units to *Attention* and *Right Shoulder Arms,* when appropriate, and face to the front. (Supplementary commands are not given in mass formations.)

(5) When all units are at *Attention*, the adjutant directs **SOUND ADJUTANT'SCALL**.

(6) The band sounds *Adjutant's Call*, and (without pause) begins playing a march. It continues until the last unit halts on the final line.

(7) Immediately after *Adjutant's Call*, the following events take place simultaneously:

*(a) All unit commanders immediately command **GUIDE ON LINE**. The guide of each unit double-times (*Port Arms*) to his position on the final line of markers, halts with his right foot on the marker, executes *Order Arms,* and then faces to the right.

(b) The adjutant aligns the guides (if necessary) and immediately marches to a position centered on the command and halfway between the post of the commander of troops and the final line. He halts and faces to the left, facing the line of troops.

(c) As soon as the guides are on line, the unit commanders (in sequence from right to left) command ***Forward,*** **MARCH**. As they approach the positions of the guides, commanders command ***Mark Time,*** **MARCH** so that the rank of squad leaders begins to *mark time* and is on line with the left shoulder of the guide. The commander allows the unit to mark time for about eight counts and then commands ***Company (Battalion),*** **HALT.** On the command of execution **HALT**, the guide executes left face, which places his right foot on the marker. If the unit is at *Right Shoulder Arms,* the unit commanders command ***Order,*** **ARMS**. The unit is then aligned by the command ***At close interval, Dress Right,*** **DRESS.** At the command of execution **DRESS**, the platoon leader of the right platoon moves by the most direct route to the right flank and verifies the alignment of as many ranks as necessary to ensure proper alignment. When he has completed verification, he returns to a position directly in front of his third squad leader, halts parallel to the formation, and faces to the right. Other platoon leaders, on the command of execution **DRESS**, position themselves directly in line with their third squad by executing one (15-inch) step to the right. The commander then commands ***Ready,*** **FRONT** and, immediately, ***Parade,*** **REST**. He faces about and assumes *Parade Rest*. For larger reviews, the command **AT EASE** may be substituted for **PARADE REST** throughout the ceremony.

NOTE: If a ceremony is conducted with a company formed in line with platoons in column, the commands **COVER** and **RECOVER** are given to align the platoons.

(8) When all units are on the final line and are at *Parade Rest,* the adjutant directs **BRING YOUR UNITS TO ATTENTION**.

(9) Unit commanders face about and (in sequence starting with the right flank unit) command ***Company (Battalion),*** **ATTENTION**. For larger formations, the commands may start with the center or right center unit, working toward both flanks. The unit commander then faces about. (10) When all units are at *Attention,* the adjutant faces about. That is the signal for the commander of troops and his staff to move from their positions near the reviewing stand to their posts midway between the line of troops and the reviewing stand and face the line of troops.

(11) When the commander of troops has halted at his post, the adjutant faces about and directs **BRING YOUR UNITS TO PRESENT ARMS**.

(12) Unit commanders face about and in sequence command ***Present,* ARMS**. They then face about and *Salute.*

(13) After all units are at *Present Arms,* the adjutant faces about, *Salutes,* and reports ***"Sir, the command is formed."***

(14) The commander of troops returns the *Salute* of the adjutant and directs **TAKE YOUR POST**. (The members of the staff do not salute.) The adjutant takes his post by facing to the half left in marching, marches forward, halts at normal interval to the right of the right flank staff member, and faces about. When the adjutant is in position, the commander of troops directs **BRING YOUR UNITS TO ORDER ARMS**. Unit commanders terminate their *Salutes;* face about; command ***Order,* ARMS**; and then face about. When all units are at *Order Arms,* the left flank staff officer commands ***Right,* FACE*;* *Forward,* MARCH; *Column Left,* MARCH; *Column Left,* MARCH; *Staff* HALT**; and ***Left,* FACE**. At that time, the staff should be centered on, and two steps in front of, the commander of troops (Figure 10-5).

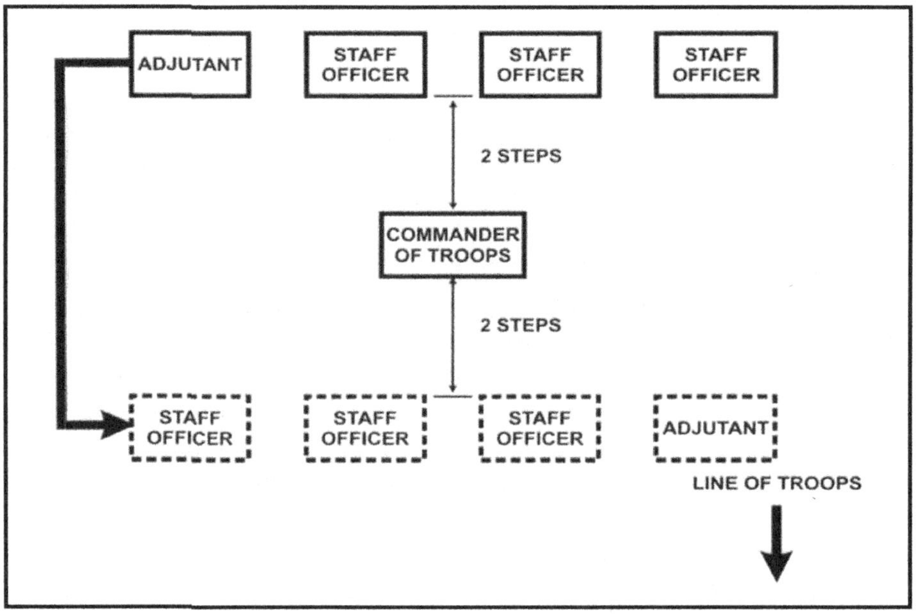

Figure 10-5. Movement of the staff to the rear of the commander of troops.

(15) When the units have completed *Order Arms* and the staff has reversed, the commander of troops then faces the reviewing stand and waits for the reviewing officer to take his post. If, for some reason, the reviewing officer has been delayed, or if a delay is prearranged, the commander of troops directs the units to ***Parade,*** **REST** and then directs them to **ATTENTION** before the reviewing officer's arrival.

b. **Presentation and Honors.** Use the following procedures to execute presentation and honors during a review.

(1) When the reviewing officer has halted at his post, the commander of troops faces about and directs **BRING YOUR UNITS TO PRESENT ARMS**. Unit commanders face about; command ***Present,*** **ARMS**; face to the front; and salute. When the units have completed this movement, the commander of troops faces about and commands his staff and himself to ***Present,*** **ARMS**.

NOTES:
1. If the reviewing officer is entitled to honors (AR 600-25), the commander of troops should verify that the salute battery (executive officer's right arm is raised) and the band (band master's arms are at the ready-play position) are prepared to render honors before he faces the reviewing officer. The *Salute* of the commander of troops is the signal for the band and salute battery to render honors. The reviewing party, and all military personnel within sight or hearing, salute on the first note of the music and terminate their *Salute* when honors have been completed.
2. If the reviewing officer is not entitled to honors, only the reviewing officer returns the *Salute* of the commander of troops.
3. During a review, *Salutes* (honors) are directed to the reviewing officer's post. In some reviews, it is appropriate for a person (colonel) not entitled honors to participate as the reviewing officer, and a person (general) entitled honors to participate as host or distinguished guest. In that situation, honors (ruffles and flourishes and cannon salute) are not appropriate.

(2) Upon completion of the presentation and or honors, the commander of troops commands his staff to ***Order,*** **ARMS**; faces about; and directs **BRING YOUR UNITS TO ORDER ARMS AND PARADE REST.**

(3) Unit commanders terminate their *Salutes,* face about, and command ***Order,*** **ARMS** and ***Parade,*** **REST**. They then face to the front and execute *Parade Rest.*

(4) When all troops are at *Parade Rest,* the commander of troops faces about. He and his staff remain at *Attention.*

NOTE: When scheduled, retreat is integrated into the review at this point. Procedures are outlined in paragraph 10-6.

c. **Inspection.** Use the following procedures to conduct an inspection during a review.

NOTE: The inspection may be omitted for decorations, awards, or retirement ceremonies.

(1) When the commander of troops has faced the reviewing stand, the reviewing officer and the host or host commander moves forward and halts three steps in front of the commander of troops. Staffs, aides-de-camp, flag bearers, and orderlies do not normally accompany the inspecting party.

(2) The commander of troops salutes and reports ***"Sir, the command is prepared for inspection."*** The band begins to play marching music on the *Salute* of the commander of troops and continues to play until the reviewing officer has returned to his post.

(3) The commander of troops guides the reviewing party to the right flank of the band. The commander of troops and the host or host commander march to the right of the reviewing officer. Whether marching or riding, the reviewing party passes between the line of brigade commanders and staffs, battalion commanders, and company commanders in division or similar-size reviews. In reviews for a battalion, or units of similar size, the inspecting party passes between the front rank of troops and the line of company commanders, or they pass immediately in front of the platoon leaders when companies are in a line formation.

(4) When the inspection is made by motor vehicle, the reviewing party enters the vehicle in front of the reviewing stand. The reviewing officer enters first and occupies the left rear position. The host or host commander occupies the right rear position. The vehicle, approaching from the left of the commander of troops, proceeds to the post of the commander of troops and stops. The commander of troops salutes, reports, and enters the vehicle, occupying the right front position.

(5) Upon the departure of the reviewing party, the senior staff officer commands the staff to ***Parade,*** **REST.** After the inspection has been completed, the senior staff officer commands ***Staff,*** **ATTENTION** before the return of the commander of troops.

(6) When the reviewing party approaches the right flank of each unit, the unit commander faces about and commands ***Company,*** **ATTENTION.** He then faces to the front and over his right shoulder commands ***Eyes,*** **RIGHT.** On the command ***Eyes,*** **RIGHT,** the company commander, executive officer, and platoon leaders execute *Eyes Right* and *Salute.* The guidon bearers execute *Eyes Right* and *Present Guidons.* Each soldier turns his head to the right and, as the reviewing officer comes into his line of vision, each soldier follows the officer with his head and eyes until the reviewing officer reaches the front. At that point, the head and eyes of each soldier remain to the front. As soon as the reviewing officer has cleared the unit, the commander commands ***Order,*** **ARMS.** He then faces about and commands ***Parade,*** **REST**, assumes the position himself, and remains facing his unit. The left flank unit remains at *Attention* until the reviewing party has cleared the right rear of the unit. Other commanders remain facing to the rear and command their units to **ATTENTION** as the party passes to the rear of their units. They then command ***Parade,*** **REST,** face to the front, and assume *Parade Rest.*

NOTES:

1. Platoon leaders give the command ***Eyes,*** **RIGHT** and *Salute* when the company is in a line formation. The company commander faces his unit, but neither he nor his guidon bearer salute.
2. When in battalion mass formations, the battalion commander faces about and commands ***Battalion,*** **ATTENTION** and ***Eyes,*** **RIGHT**, but neither he nor his staff salute or execute *Eyes Right.* The unit commanders,

executive officers, platoon leaders, and guidon bearers salute. The battalion commander remains facing his unit, until the reviewing officer has passed the left front of his unit, at which time he commands ***Order,*** **ARMS** and ***Parade,*** **REST**. He then commands his staff to ***Parade*** **REST** but remains facing his unit.

(7) As the reviewing party approaches the Colors, the commander of troops inconspicuously commands ***Present,*** **ARMS** (six steps from the Colors) and ***Order,*** **ARMS** (six steps beyond the Colors). They do not salute when passing to the rear of the Colors. The Color guard and bearers execute *Eyes Right.* The organizational Color dips (salutes).

(8) After passing in front of the troops (to include the salute battery), the inspection continues along the rear of the troops and terminates at the right flank of the band. The commander of troops commands ***Party,*** **HALT**. The bandmaster has the band play softly until the reviewing party members begin marching back to their posts. The commander of troops faces to the half left in marching, takes two steps, halts, and faces about. The host or host commander repositions himself to the left of the reviewing officer. When the host or host commander is in position, the commander of troops and the reviewing officer exchange *Salutes.* Upon termination of the *Salutes,* the reviewing officer and host or host commander immediately face to the half left in marching and return to their posts. The commander of troops hesitates momentarily and then faces to the right in marching and returns to his post (Figure 10-6, page 10-14).

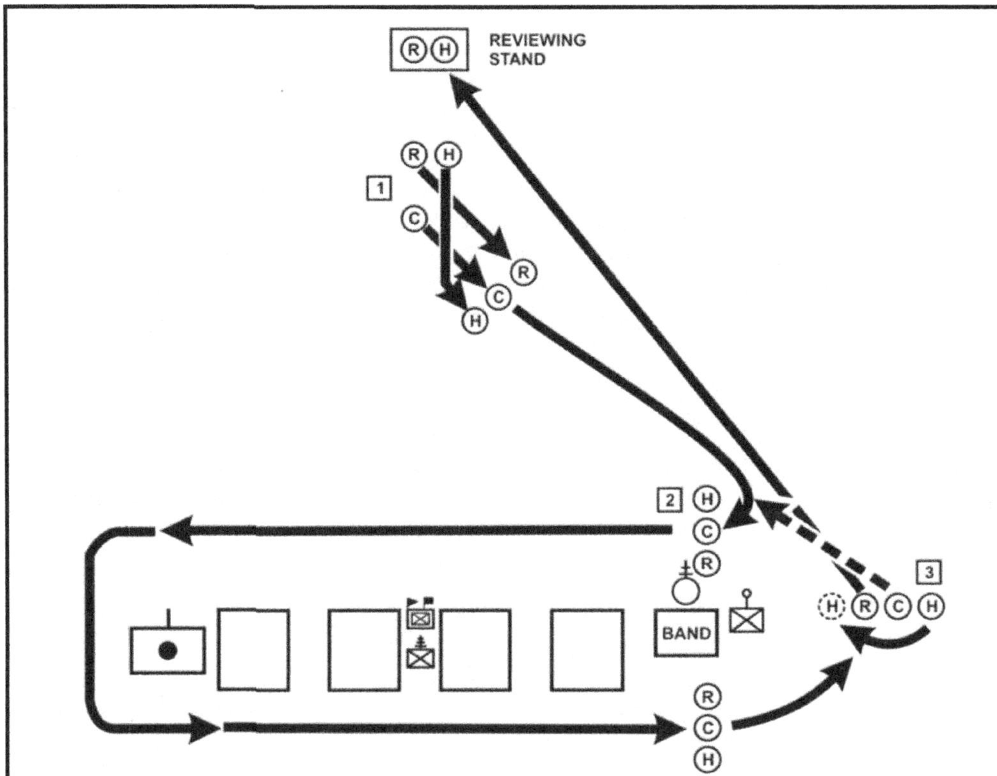

1 COMMANDER OF TROOPS REPORTS TO THE REVIEWING OFFICER AND FORMS REVIEWING PARTY.

2 REVIEWING PARTY BEGINS INSPECTION OF TROOPS.

3 REVIEWING PARTY TERMINATES INSPECTION; HOST/HOST COMMANDER POSITIONS HIMSELF TO THE LEFT OF REVIEWING OFFICER AND THEY RETURN TO REVIEWING STAND; COMMANDER OF TROOPS MOVES INTO POSITION TO RETURN TO HIS POST IN FRONT OF THE FORMATION.

LEGEND

Ⓡ REVIEWING OFFICER

Ⓗ HOST/ HOST COMMANDER

Ⓒ COMMANDER OF TROOPS

NOTE: THE HOST/HOST COMMANDER AND COMMANDER OF TROOPS WILL HESITATE WHEN NECESSARY TO ALLOW THE REVIEWING OFFICER TO PRECEDE THEM.

WHEN A DISTINGUISHED GUEST PARTICIPATES, HE MARCHES ON THE RIGHT OF THE HOST/HOST COMMANDER DURING THE INSPECTION: OR THE HOST/HOST COMMANDER AND DISTINGUISHED GUEST MAY MARCH TO THE REAR OF THE REVIEWING OFFICER AND COMMANDER OF TROOPS.

Figure 10-6. Movement of the reviewing party.

d. **Honors to the Nation.** Use the following procedures to execute honors to the Nation when conducting a review.

(1) When the reviewing party members have returned to their posts, the commander of troops faces about and directs **BRING YOUR UNITS TO ATTENTION**. After all units are at *Attention*, the commander of troops gives the command ***Colors Center*** (pause), **MARCH**. On the preparatory command ***Colors Center***, the staff faces to the right. On the command of execution **MARCH,** the staff begins marching forward to a position offset to the left front of the reviewing stand, halts, and faces to the left. The

Colors take seven steps forward and halt. The commander of troops marches forward until he is three steps in front of the color detail, halts, and then faces about.

*(2) To move the Colors forward, the commander of troops commands ***Colors Forward,*** **MARCH**. On the command of execution **MARCH,** the commander of troops and the color detail march forward and then halt when the commander of troops has reached his original post. The commander of troops executes the *Hand Salute* and reports to the host commander, ***"Sir, the Colors are present."*** The host commander then returns the *Salute* and directs the commander of troops **ASSUME YOUR POST AND RESENT THE COMMAND.** On this directive the commander of troops faces to the right in marching and marches by the most direct route to his post two steps in front of and centered on his staff, halts perpendicular to his staff, then faces to the right. After facing to the right, the commander of troops directs **BRING YOUR UNITS TO PRESENT ARMS.**

*(3) Unit commanders face about and command ***Company (Battalion), Present,*** **ARMS**. Each commander then faces about and salutes. When all units have completed these movements, the commander of troops faces about and commands ***Present,*** **ARMS** for himself, his staff, and the Colors detail. On the execution of the *Hand Salute* by the commander of troops, the band begins to play the National Anthem.

(4) The reviewing party and all military spectators salute while the National Anthem is being played (Appendix K).

(5) Upon completion of the National Anthem, the commander of troops, his staff, and the Colors detail terminate their *Salutes* on his command. He then faces about and directs **BRING YOUR UNITS TO ORDER ARMS AND PARADE REST.**

(6) Unit commanders terminate their *Salutes;* face about; command ***Order,*** **ARMS** and ***Parade,*** **REST**; face about; and execute *Parade Rest.*

(7) When the last unit has executed the directive, the commander of troops faces about and commands his staff and himself to ***Parade,*** **REST**.

NOTE: Other ceremonial activities may be integrated into the review at this point. The sequence of events and actions of individuals are discussed in paragraphs10-4 and 10-5.

e. **Remarks.** After the commander of troops and his staff have assumed *Parade Rest,* the reviewing officer, the host or host commander, or the distinguished guest may address the command.

(1) Upon completion of the remarks, the commander of troops brings his staff to *Attention,* faces about, and directs **BRING YOUR UNITS TO ATTENTION**. Unit commanders face about and command ***Company (Battalion),*** **ATTENTION**, and then face about. The commander of troops faces about and commands ***Detachment, POST*** (pause), **MARCH**. On the command **POST**, the Colors *Reverse March* and halt.

(2) On the command **MARCH**, Colors step off, return to their original posts, and the band begins to play.

(3) As the Colors pass his position, the commander of troops and his staff face to the left. He marches his staff back to the center of the field and faces them to the right as he faces to the left. After the Colors are in position, the commander of troops then faces the reviewing officer.

f. **March in Review.** Use the following procedures to conduct a march in review.

(1) Upon completion of the remarks, the commander of troops commands his staff to **ATTENTION**, faces about, and directs **BRING YOUR UNITS TO ATTENTION**. Unit commanders face about and command ***Company (Battalion),*** **ATTENTION** and face back to the front. When the units are at *Attention,* the commander of troops faces the reviewing officer.

(2) When the commander of troops has faced the reviewing officer, the host or host commander directs **PASS IN REVIEW**.

(3) The commander of troops faces about and directs **PASS IN REVIEW**. The band is then faced to the right and marched to a position that enables it to move straight forward onto the line of march without an initial turning movement, halts, and faces to the left. The left turn marker should be positioned to minimize the movement of the band (Figure 10-7).

NOTE: The commander takes no further action until the drum major turns his head and eyes and points his arm toward the commander.

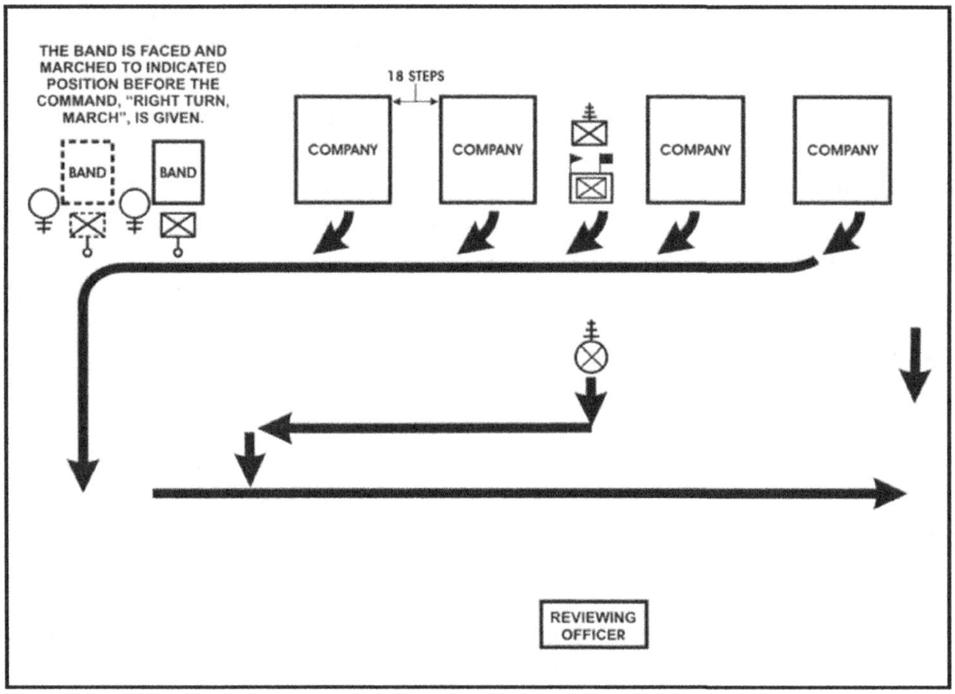

Figure 10-7. March in review in column with units in mass.

(4) Unit commanders face to the right when the commander of troops directs **PASS IN REVIEW**. If weapons are to be carried at *Right Shoulder Arms,* the commander faces about and commands ***Right Shoulder, ARMS.*** He then faces to the left. The commander next to the band waits until the band moves into position. He then commands ***Right turn,*** **MARCH** loud enough for the band to hear. The band takes this command as its signal to begin playing and to march forward onto the line of march.

(5) Other units move out in procession in the same manner and follow in column at the prescribed distance (Figure 10-8). For larger reviews, commanders may command their units (in sequence) to *Parade Rest* while waiting their turn to move onto the line of march.

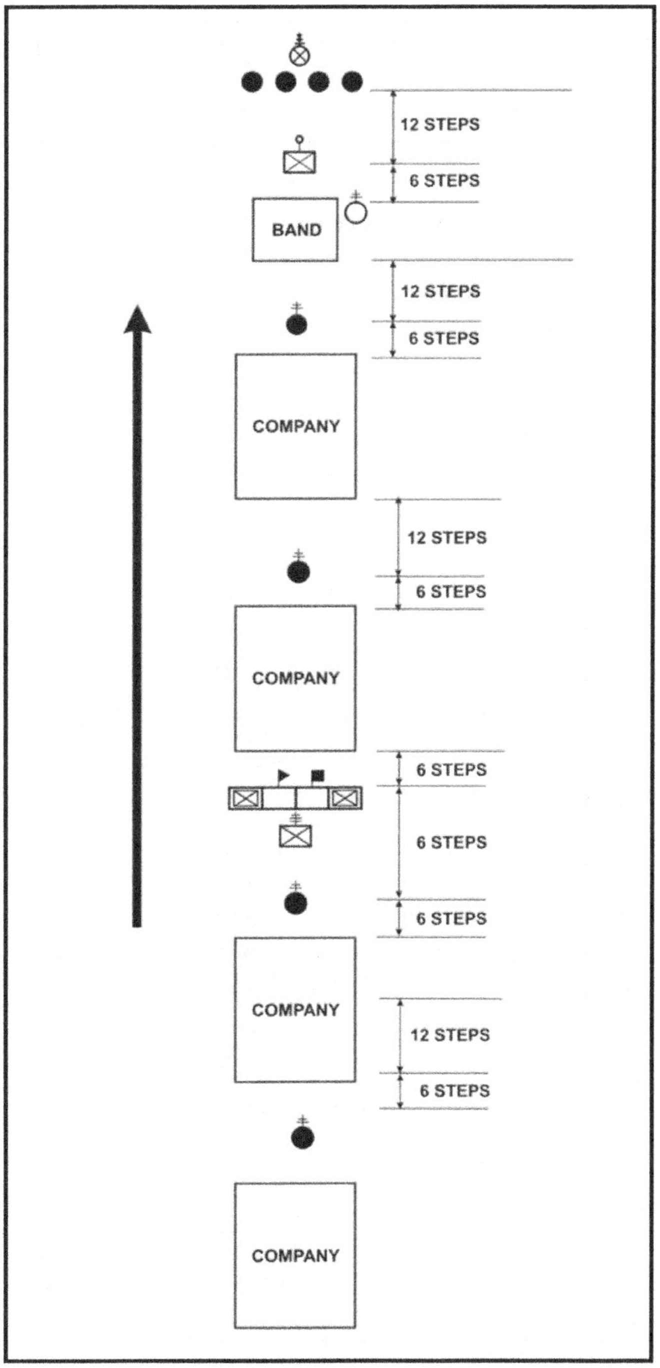

Figure 10-8. Battalion in column with companies in mass.

(6) The band and each following unit change direction at points indicated by markers. The commander commands, ***Left turn,*** **MARCH.** The commander faces about while marching as his unit is making the turn. When his unit has completed the turn, the commander commands, ***Forward*** **MARCH,** and faces back to the front.

(7) All commanders, except the commander of troops, move with their staff into positions in the column and at the head of their respective units just before turning onto the reviewing line (Figure 10-9).

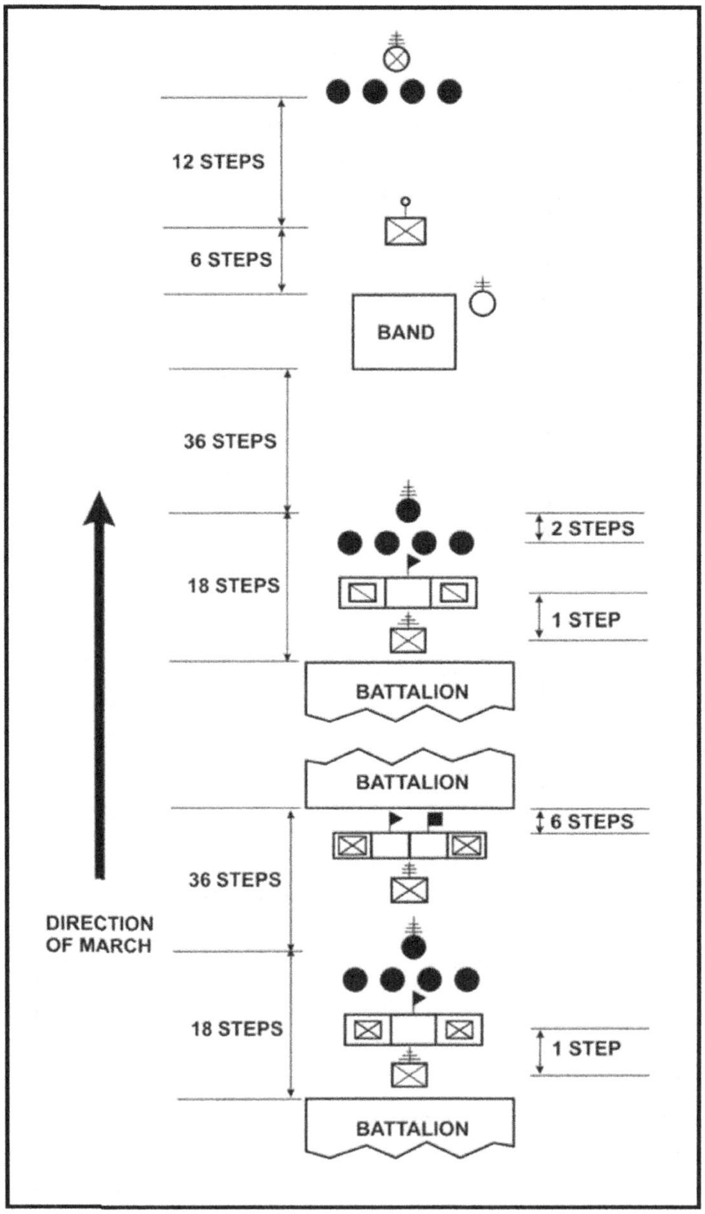

Figure 10-9. Brigade in column with battalions in mass.

(8) The commander of troops and his staff move forward and execute turning movements to arrive at a position 12 steps in front of the drum major on the reviewing line.

(9) On command, the commander of troops, the brigade and battalion commanders, their staffs, and the command sergeant major execute ***Eyes,*** **RIGHT** and salute at the *Eyes Right* marker. The commander commands ***Ready,*** **FRONT** and terminates the *Salute* when the staffs have reached the *Ready Front* marker.

(10) The reviewing officer returns only the *Salute* of the commander of troops. (The return of the *Salute* by the reviewing officer represents the *Salute* for all subordinate commanders. This enables the reviewing officer to observe the review without being interrupted by frequent *Salutes.)* The reviewing officer, the host or host commander, their staffs, and military spectators salute the National Color when it passes.

(11) After terminating the *Salute,* the commander of troops and his staff (without command) execute three wheeling movements and take their post with the commander of troops on line with and to the right of the reviewing officer.

(12) Troop units execute *Eyes Right* on command from their company commander or from the center company commander when in battalion mass. Commanders give the preparatory command ***Eyes*** over their right shoulder two steps from the marker as the right foot strikes the marching surface. The command of execution **RIGHT** is given when the right foot strikes the marching surface again and on line with the marker. On the preparatory command, the guidon bearer executes *Raised Guidon.* On the command of execution, the company commander, executive officer, and platoon leaders execute *Eyes Right* and the *Hand Salute.* The guidon bearer executes *Eyes Right* and *Present Guidon.* The company first sergeant only executes *Eyes Right.* The right file continues to look straight forward and maintains correct distance. All other members execute *Eyes Right* and maintain alignment. When the rear of the unit has passed six steps beyond the reviewing officer, company commanders command ***Ready*** as the left foot strikes the marching surface and **FRONT** the next time the left foot strikes the marching surface. Unit personnel end their *Salutes* and turn their heads and eyes to the front. The guidon bearer executes *Raised Guidon* on the command ***Ready*** and returns to the *Carry* position on the command **FRONT**.

(13) When passing the reviewing officer, the bandmaster salutes and continues to look straight forward. Simultaneously, the drum major executes *Eyes Right* and salutes. The other members of the band continue to play marching music without interruption. When the band has passed the reviewing officer, the drum major has the band execute three left turns into a position in front of and facing the reviewing officer and at least 12 steps from the left flank of the marching troops. As the Colors pass, the bandmaster and drum major salute while the band continues to play marching music without interruption.

(14) As the Color guard passes the reviewing officer, each member, except the right flank man, executes *Eyes Right* on the command of the senior Color sergeant. The organizational color is dipped in *Salute.*

g. **Conclusion.** Follow these procedures to conclude a review.

(1) When the last troop element has passed the reviewing stand and has executed *Ready Front,* the band goes into an eight-bar drum cadence. On the first beat following the drum cadence, the band begins playing (in place) the official version of "The Army

Goes Rolling Along," beginning at the introduction. At the appropriate time, the drum major has the band march forward and execute a left turn onto the reviewing line while continuing to play until the completion of the chorus.

NOTE: The band may play music representative of the command after the last troop element has passed the reviewing stand and before playing "The Army Goes Rolling Along."

(2) At the completion of the Army song, the commander of troops and the reviewing officer face each other and exchange *Salutes*, thus officially terminating the ceremony.

(3) It is appropriate for spectators to rise, stand at *Attention*, and sing while "The Army Goes Rolling Along" is being played.

10-4. REVIEW WITH DECORATIONS, AWARDS, AND INDIVIDUAL RETIREMENT (SEQUENCE OF EVENTS)

Unless otherwise specified, the sequence of events for a review with decorations, awards, and individual retirement is the same as described in paragraph 10-3. Only changes to the sequence of events are listed herein.

NOTE: The procedures mentioned provide latitude for recognition of one or a group of retirees of various grades. However, the host or host commander should consider it appropriate to have the review conducted in the retiree's honor when only one distinguished individual is retiring and allow that individual to participate as the reviewing officer. When the Colors halt, the reviewing officer (retiree) positions himself in front of and centered on the Colors facing the reviewing stand. After awards are presented by the host, the retiree returns to the reviewing officer's post. If there are several awardees and or retirees, it may be desirable to have the host or host commander act as the reviewing officer. In such case, the following sequence will apply.

a. **Formation of Troops.** Formation of troops is executed the same as described in paragraph 10-3.

b. **Presentation and Honors.** Presentation and honors is executed the same as described in paragraph 10-3.

NOTE: When scheduled, retreat is integrated into the review at this point. Procedures are outlined in paragraph 10-6.

c. **Inspection.** The inspection may be omitted for decorations, awards, or retirement ceremonies. If the inspection is omitted, the commander of troops (following *Order Arms* after the presentation and or honor, or after "To the Color" is played) commands ***Persons to be honored and Colors center*** (pause), **MARCH.** The ceremony then continues (as described in this paragraph). If the inspection is **not** omitted, use the procedures described in paragraph 10-3c to conduct an inspection during a review.

d. **Honors to the Nation.** (Omitted if retreat is conducted as part of the review.)

NOTE: Other ceremonial activities may be integrated into the review at this point.

e. **Remarks.** After the commander of troops and his staff have assumed *Parade Rest,* the reviewing officer, the host or host commander, or the distinguished guest may address the command.

(1) Upon completion of the remarks, the commander of troops brings his staff to *Attention,* faces about, and directs **BRING YOUR UNITS TO ATTENTION**. Unit commanders face about and command ***Company (Battalion),*** **ATTENTION**, and then face about. The commander of troops faces about and commands ***Detachment, POST*** (pause), **MARCH**. On the command **POST**, the following actions occur simultaneously:

(a) Colors *Reverse March* and halt.

(b) Awardees and or retirees execute a *Right Face;* guidon bearers and or commanders execute an *About Face.*

(2) On the command **MARCH**, Colors and awardees step off and the band begins to play.

(a) Persons who were decorated march forward, execute two *Column Lefts,* halt on line (six steps to the left of the reviewing officer), and execute a *Left Face.* The commands **HALT** and ***Left,*** **FACE** are given by the last man in the file.

(b) Colors step off and return to their original posts.

(3) As the Colors pass his position, the commander of troops faces himself and his staff to the left, marches his staff back to the center of the field, and faces them to the right. The commander of troops faces to the left; when the Colors are in position he then faces the reviewing officer.

f. **March in Review.** March in review is executed the same as described in paragraph 10-3.

g. **Conclusion.** Conclusion is executed the same as described in paragraph 10-3.

10-5. REVIEW WITH CHANGE OF COMMAND, ACTIVATION, OR DEACTIVATION (SEQUENCE OF EVENTS)

Unless otherwise specified, the sequence of events for a review with change of command, activation, or deactivation is the same as described in paragraph 10-3. Only changes to the sequence of events are listed herein.

a. **Formation of Troops.** Formation of troops is executed the same as described in paragraph 10-3.

b. **Presentation and Honors.** Presentation and honors is executed the same as described in paragraph 10-3.

c. **Inspection.** The inspection may be omitted for change of command, activation, or deactivation ceremonies. If the inspection is **not** omitted, use the procedures described in paragraph 10-3c to conduct an inspection during a review. When the reviewing officer has completed the inspection, the commander of troops faces about and directs **BRING YOUR UNITS TO ATTENTION**. Unit commanders face about and command ***Company (Battalion),*** **ATTENTION**, then they face about. When the command has completed the movement, the commander of troops commands ***Colors Center*** (pause), **MARCH**. On that command, the Colors are brought forward in the same manner as described in paragraph 10-3d.

d. **Honors to the Nation.** (Omitted if retreat is conducted as part of the review.) On the completion of honors to the nation, the commander of troops commands ***Detachment order,*** **ARMS**, faces about, and directs **BRING YOUR UNITS TO ORDER ARMS AND PARADE REST**. Unit commanders terminate their *Salutes,* face about and command ***Order,*** **ARMS** and ***Parade,*** **REST**. They then face about and assume *Parade Rest.* The commander of troops faces about and commands ***Parade*** **REST**; the commander of troops and his staff execute *Parade Rest.*

NOTE: Other ceremonial activities may be integrated into the review at this point.

e. **Remarks.** As soon as the commander of troops and his staff have executed *Parade Rest,* the reviewing party moves forward to within four steps of the Colors and halts. As the party marches forward, the senior commander positions himself between the old and the new commander. The command sergeant major moves from his post (at the rear of the reviewing party) by the most direct route and halts directly in front of the organizational Color.

NOTE: At the beginning of the ceremony, the outgoing commander should assume the role of host or host commander. The senior official or commander designated to "pass the organizational Color" should assume the role of an honored guest or reviewing officer except during that brief portion of the ceremony wherein he may be called upon to act as a host or host commander for the purpose of passing the organizational Color. Since the inspection of troops normally occurs early in the ceremony, the incoming commander is also at that time treated as a guest, and is expected to accompany the outgoing commander and senior official during the inspection. After the organizational Color has been passed, the incoming commander should assume the role of host or host commander, and the senior official and outgoing commander both become honored guests or reviewing officers at this point and remain as such for the balance of the ceremony (Figure 10-10).

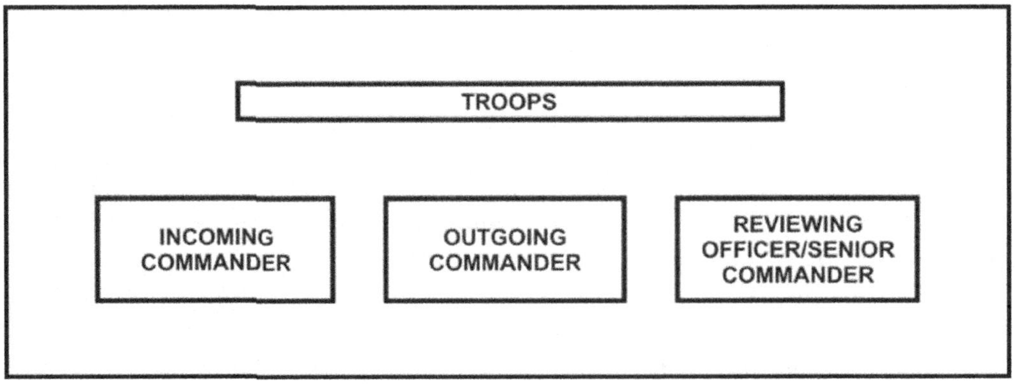

Figure 10-10. Reviewing party for change of command.

(1) ***Change-of-Command Ceremonies.*** The outgoing commander positions himself four steps in front of his organizational Color; the senior commander is directly to his left. The old and new commanders then take one step forward and execute *Facing* movements so that they are facing each other.

(a) The command sergeant major removes the organizational Color from the Color bearer's sling (with his right hand above his left hand), and faces about. The narrator reads the assumption-of-command order.

(b) Upon completion of the reading, the command sergeant major steps forward and presents the organizational Color to the outgoing commander (1, Figure 10-11), who grasps the organizational Color with his left hand above his right hand. The outgoing commander passes the organizational Color to the senior commander (2, Figure 10-11), who grasps the Color with his right hand above his left hand and, in turn, passes the organizational color to the new commander (3, Figure 10-11), who grasps the organizational color with his left hand above his right hand. The incoming commander passes the organizational color to the command sergeant major (4, Figure 10-11) who grasps it with his right hand above his left hand, faces about and returns the organizational Color to the Color bearer's sling.

(c) As the command sergeant major faces about, both commanders then execute *Facing* movements back to their original direction facing the Colors. The command sergeant major and the reviewing party face about and return to their post. At this point, the commander makes his remarks.

NOTES:
1. This procedure allows the organizational Color to be over the heart of the incoming and outgoing commanders.
2. For a company level change of command, the reviewing party is lined up the same as in larger units. The guidon takes the place of the organizational Color and the first sergeant replaces the command sergeant major. The guidon is passed in the same manner as the organizational Color. The sequence of events remains the same with modifications made to fit available assets.

(d) Upon completion of the remarks, the commander of troops brings his staff to *Attention,* faces about, and directs **BRING YOUR UNITS TO ATTENTION**. Unit commanders face about and command ***Company (Battalion),*** **ATTENTION**, and then *face about.* The commander of troops faces about and commands ***Detachment, POST*** (pause), **MARCH**. On the command **POST**, the following actions occur simultaneously:

- Colors *Reverse March* and *Halt.*
- Awardees and or retirees execute a *Right Face;* guidon bearers and or commanders execute an *About Face.*

(e) On the command **MARCH**, Colors and awardees (if present) step off and the band begins to play.

- Persons who were decorated march forward, execute two *Column Lefts,* halt on line (six steps to the left of the reviewing officer), and execute a *Left Face.* The commands **HALT** and ***Left,*** **FACE** are given by the last man in the file.
- Colors step off and return to their original posts.

- As the Colors pass his position, the commander of troops and his staff face to the left. He marches his staff back to the center of the field and faces them to the right as he faces to the left. After the Colors are in position, the commander of troops then faces the reviewing officer.

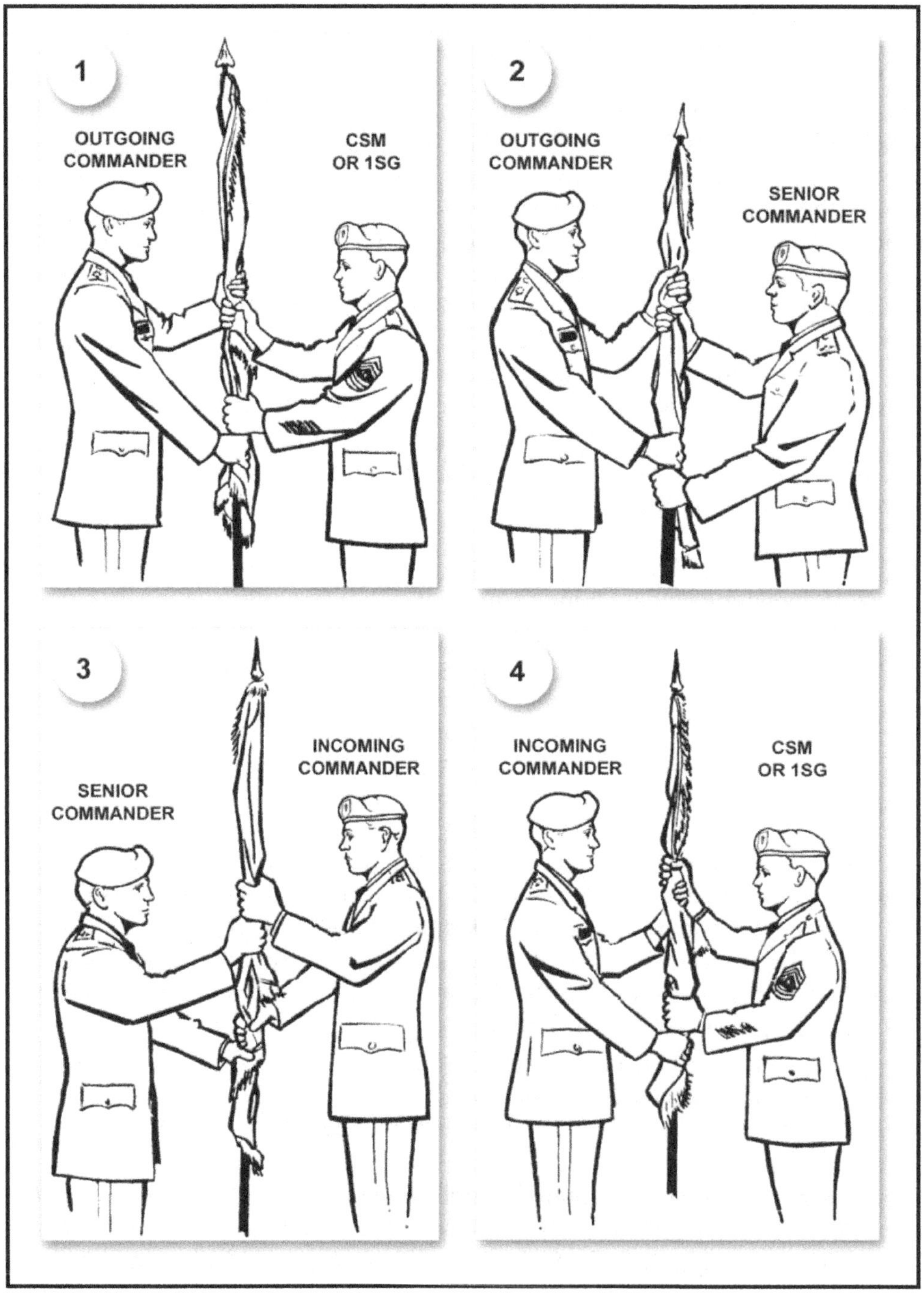

Figure 10-11. Passing of organizational Color for change of command.

(2) ***Activation Ceremonies***. The position of the senior commander is four steps in front of the Color bearer (with an empty sling for the organizational Color of the unit to be activated) with the unit commander to his left. The command sergeant major moves from his post with the cased organizational Color and positions himself in front of and facing the senior commander. The narrator reads the activation order.

(a) Upon completion of the reading, the command sergeant major presents the cased organizational Color to the senior commander and steps backward. The senior commander lowers the cased organizational Color to allow the command sergeant major to uncase it. The command sergeant major folds the case and places it inside the Color bearer's cartridge belt (rear center).

(b) With the organizational Color uncased, the senior commander rotates the staff to allow the organizational Color to unfurl and drape freely. The senior commander presents the organizational Color to the commander of the unit to be activated. The command sergeant major steps forward and accepts the organizational Color from the unit commander and places it in the Color bearer's sling.

(c) When the command sergeant major faces about, the reviewing party faces about and returns to the reviewing stand. The command sergeant major returns to his original post. Procedures from this point forward are the same as for change-of-command ceremonies.

(3) ***Deactivation Ceremonies***. The position of the commander is four steps in front of the organizational Color. The command sergeant major moves from his post and positions himself in front of the organizational Color. He moves the organizational Color from the bearer's sling and faces about. The narrator reads the deactivation order.

(a) Upon completion of the reading, the command sergeant major presents the organizational Color to the commander and steps backward. The organizational Color is grasped by the commanders and the command sergeant major as described for change-of-command ceremonies. The commander rotates and lowers the organizational Color, allowing the command sergeant major to case the organizational Color.

(b) Upon completion of the casing, the commander presents the cased organizational Color to the senior commander and the command sergeant major steps forward and accepts the organizational Color from the senior commander. The reviewing party faces about and returns to the reviewing stand.

(c) The command sergeant major hands the cased organizational Color to a designated soldier who marches from the field and the command sergeant major returns to his post. The Color bearer of the deactivated unit marches with the Color guard for the remainder of the ceremony with an empty sling. Procedures from this point forward are the same as for change-of-command ceremonies.

10-6. REVIEW WITH RETREAT (SEQUENCE OF EVENTS)

Unless otherwise specified, the sequence of events and individual actions for a review with retreat are the same as described in paragraph 10-3. Only changes to the sequence of events are listed herein.

NOTE: Honors to the nation is omitted when retreat is conducted as part of a review.

a. **Formation of Troops.** Formation of troops is executed the same as described in paragraph 10-3.

b. **Presentation and Honors.** Presentation and honors is executed the same as described in paragraph 10-3.

c. **Retreat.** Upon completion of the presentation and or honors, the commander of troops commands his staff to ***Order,*** **ARMS**; faces about; and directs **BRING YOUR UNITS TO ORDER ARMS AND PARADE REST.** Unit commanders terminate their *Salutes,* face about, and command ***Order,*** **ARMS** and ***Parade,*** **REST**. They then face to the front and execute *Parade Rest.* The commander of troops directs **SOUND RETREAT**, faces about, and commands his staff to ***Parade,*** **REST**. As soon as the commander of troops and his staff are at *Parade Rest*, the band sounds retreat. At the conclusion of retreat, the commander of troops commands his staff to **ATTENTION**, faces the troops, and directs **BRING YOUR UNITS TO ATTENTION AND PRESENT ARMS**. When the units have completed these movements, the commander then faces the reviewing officer and commands his staff to ***Present,*** **ARMS**. This is the signal for the band to play "To the Color."

(1) When the ceremony is held on an Army post, the cannon is fired on the last note of "Retreat," and the flag is lowered while "To the Color" is played. (See Appendix F regarding appropriate procedures.)

(2) The reviewing officer and his staff stand at *Attention* during the sounding of the retreat. On the first note of "To the Color," they salute and hold the *Salute* until the last note is played.

(3) On the last note of the music, the commander of troops commands ***Order,*** **ARMS** for himself and his staff, faces about, and directs **BRING YOUR UNITS TO ORDER ARMS AND PARADE REST**. When the units have completed these movements, the commander of troops then faces the reviewing officer.

(4) The review then proceeds with the inspection as in a normal review; if omitted, other parts are integrated into the review as described in paragraphs 10-4 and 10-5.

10-7. SPECIAL REVIEW

A special review is conducted by a composite or representative unit and serves the same purposes as a standard review when a larger formation is not practical. The composite unit normally consists of a band, Colors, two or more platoons, and a salute battery (when appropriate). The reduced formation is not a reason to eliminate any of the components of a review or change their sequence.

a. A special review is essentially the same as a battalion or larger unit review, except:

(1) The troops are formed on a final line.

(2) *Attention* and *Adjutant's Call* are not sounded.

(3) Neither the adjutant nor a staff participate.

(4) Unit guidons are not used in this formation. (Organizational Colors represent the unit conducting the review.)

(5) The commander of troops gives commands rather than directives.

(6) The band positions itself in front of the reviewing officer by executing a *Left Turn* at a point midway between the final line and the reviewing line, and then turns right when on line and centered on the reviewing officer.

(7) Platoons pass the reviewing stand in line formation rather than column formation.

(8) On the directive ***Pass in review*** the commander of troops commands ***Right,*** **FACE** (***Right Shoulder,*** **ARMS**, if appropriate); ***Forward,*** **MARCH**. The band steps off playing a march on the command of ***Forward,*** **MARCH**. Platoon leaders remain six steps to the front of and centered on their platoons. Platoon sergeants remain one step to the rear of and centered on their platoon.

(9) Platoon leaders command ***Column Left,*** **MARCH** at the first turn marker and ***Left Flank,*** **MARCH** when centered on the reviewing line.

b. Indoor ceremonies retain the same sequence as a normal review, excluding elements that are precluded because of space. Generally, a pass in review cannot be conducted indoors.

This page intentionally left blank.

CHAPTER 11
PARADES

The term "parade" had various meanings to Continental Army troops camped at Valley Forge. It could mean to form, march, and drill. Present day parade procedures originated from the daily activities conducted then to form, organize, instruct, and issue parole and countersign words to the various guards on duty (outpost, picket, camp, and quarters). Early parades also enabled commanders to give special instructions to subordinate leaders and to make command announcements. In the U.S. Army regulations of 1863, reviews were a type of parade, and "dress parades" were conducted daily, except on extraordinary and urgent occasions. The parade remains basically the same as the review except that it has retained its original intent—a method whereby unit commanders could inspect troops, present awards, and issue information. The sequence of a parade has the following steps—formation of troops, sound off by a band, honors to the nation, presentation, manual of arms, report, orders published, officers center, pass in review.

11-1. HISTORY

This paragraph discusses the history of the parade and how it applies to today's Army.

a. Battalion and brigade parades are opened by adjutant's call sounded on a trumpet. This call has opened parades of the American forces for over 150 years. After some preliminaries, the adjutant directs **SOUND OFF** and the band plays three chords. Having sounded off, the band marches in front of the troops and then countermarches to its original position. This *March* across the front of the line is said to have originated with the Crusades (A.D. 1095 to 1260). The troops offering themselves for holy service were drawn up in a long formation and the band countermarched only before those chosen to serve. In American ceremonies, the sound off has been handed down as a ceremonial tradition.

b. After the band sounds off, the commanding officer may, if he wishes, give the troops some *Facings* or movements of the manual of arms. This. was the traditional way in which he established control over his command, and it has been practiced in the United States since the revolutionary war.

c. Parades in the revolutionary army were usually held at "Troops" or about 8 o'clock in the morning and was the time used for sergeants to call the roll and report it to the adjutant.

d. After the sergeants had returned to their posts, the commander directed that necessary orders and information be reported. After the adjutant had read the orders, he announced *"Parade is dismissed."* At that time, officers moved to a position centered on the commanding officer. As the commander gave specific instructions to the officers, the first sergeants would march their units back to their respective company areas where they would be dismissed.

11-2. PREPARATION

The appearance and movement of troops in formation are the primary characteristics of a parade. The preparation and organization of troops for a parade are similar to that for a review. In a parade, since the commander of troops is also the reviewing officer, the distance between the troops and the commander is greater than that for a review. Other differences are that the band conducts *Sound Off,* the inspection is omitted, a report is rendered, orders are published, and officers and guidons are marched forward and centered on the commander. If retreat is scheduled, honors to the nation are conducted concurrently with retreat.

11-3. CEREMONIAL BATTALION PARADE

The procedures for conducting a ceremonial battalion parade are discussed herein.

a. **Formation of Troops.** Units are moved to their positions on the ready line in the most convenient manner. The commander prescribes the routes, sequence, and time of arrival on the ready line. He and his staff take their posts before adjutant's call. The procedure for moving from the ready line to the final line is the same as for a review.

b. **Sound Off.** After the battalion has been formed, aligned, and given *Parade Rest* on the final line, the adjutant directs **SOUND OFF** and assumes the position of *Parade Rest.* He remains facing the formation.

(1) At the directive *Sound Off,* the band plays three sound off chords. At the conclusion of the third chord, the band moves forward playing a march in quick time. The band executes a left turn to march across the front of the troops. At the left of the line of troops, the band countermarches and returns over the same ground to the right of the line. After the band has passed beyond the right of the troops, it executes a right turn, countermarches again, and halts in its original position.

(2) When the band has halted, it ceases playing at the next convenient place in the music and again plays the three sound off chords. (Trooping the line by the band may be eliminated when ceremonies are conducted indoors during inclement weather and space is limited).

NOTE: Retreat, when scheduled, is integrated at this point. Upon completion of the sound off chords, the band pauses briefly and then plays retreat (by the trumpet section) without a command from the adjutant. Upon completion of retreat, the adjutant comes to *Attention* and directs **BRING YOUR UNITS TO ATTENTION AND PRESENT ARMS**. The commander and his staff assume *Parade Rest, Attention, Present Arms,* and *Order Arms* with the adjutant. The adjutant faces about and salutes. The adjutant's *Salute* is the signal for the band to play "To the Color" or the National Anthem. At the conclusion of "To the Color" or the National Anthem, the adjutant comes to *Order Arms*, and faces about, and directs **BRING YOUR UNITS TO ORDER ARMS**. After the last unit comes to *Order Arms*, the adjutant faces about.

c. **Honors to the Nation.** If retreat is not conducted, the adjutant (after the band has completed *Sound Off*) assumes the position of *Attention* and directs **BRING YOUR UNITS TO ATTENTION AND PRESENT ARMS.** Unit commanders assume the position of *Attention*, face about, and command ***Company,*** **ATTENTION,** ***Present,*** **ARMS.** They face about and salute. When all units are at *Present Arms*, the adjutant faces about and salutes. This is the signal for the band to play the National Anthem. The battalion commander and his staff execute ***Present,*** **ARMS** on his command on the first note of the music.

d. **Presentation.** Upon completion of the National Anthem, the battalion commander and his staff execute ***Order,*** **ARMS** on his command. The adjutant terminates his *Salute*, faces about, and directs **BRING YOUR UNITS TO ORDER ARMS**. Unit commanders terminate their *Salute*, face about, command ***Order,*** **ARMS**, and then face (back) to the front. When all units are at *Order Arms*, the adjutant faces about, salutes, and reports ***"Sir, the parade is formed."*** The battalion commander returns the *Salute* and commands **POST.** The adjutant marches forward, passes by the battalion commander's right, and takes his post as the right flank member of the staff.

e. **Manual of Arms**. After the adjutant has joined the staff, the battalion commander commands such movements in the manual of arms as he may desire. When desired, the commander may direct, rather than command, **HAVE YOUR UNITS EXECUTE THE MANUAL OF ARMS**. The unit commanders, in sequence from right to left, command ***Right Shoulder,*** **ARMS;** ***Port,*** **ARMS;** ***Left Shoulder,*** **ARMS;** ***Order,*** **ARMS**. When the unit on the right has completed all of the movements, the next unit then begins and so on until the last unit has completed the movements.

f. **Report.** When all units have completed the manual of arms, the battalion commander then directs **RECEIVE THE REPORT**. The adjutant returns to his position at the center of the command and commands **REPORT**. Unit commanders in succession, from right to left, salute and report ***"_________ Company, all present or accounted for."*** The adjutant returns each commander's *Salute*. After receiving the report, the adjutant faces about, salutes, and reports ***"Sir, all present or accounted for."***

g. **Publishing of Orders.** The battalion commander returns the *Salute* and directs **PUBLISH THE ORDERS**. The adjutant faces about and directs **ATTENTION TO ORDERS** (he then reads the orders).

h. **Officers Center March.** After reading the orders, the adjutant directs **BRING YOUR UNITS TO PARADE REST.**

(1) Unit commanders face about and command ***Parade,*** **REST**. They then face about and come to *Parade Rest.* When all units are at *Parade Rest,* the adjutant commands (loud enough for the band to hear) ***Officers*** (pause), ***Center*** (long pause), **MARCH**. He then faces about and takes his post with the staff.

(a) On the command ***Officers,*** all officers come to *Attention* and guidon bearers come to *Carry Guidon.*

(b) On the command ***Center,*** when companies are in mass formation, the company commanders and guidon bearers face to the center. Officers commanding platoons move one step forward and face to the center. Executive officers move through the interval between units nearest to the center and take their posts in the column formed by the platoon leaders.

(c) At the command **MARCH,** the band plays marching music and continues to play until the officers have halted in front of the commander. Officers and guidon bearers close to the center, halt, and individually face to the front. Company commanders, when moving to the center, face half right or half left in marching, march to the front, and close on a line four steps in advance of the line of guidon bearers. The guidon bearers close on their own line, each taking a post to the rear of his company commander. All other officers close on the line of platoon leaders (Figure 11-1).

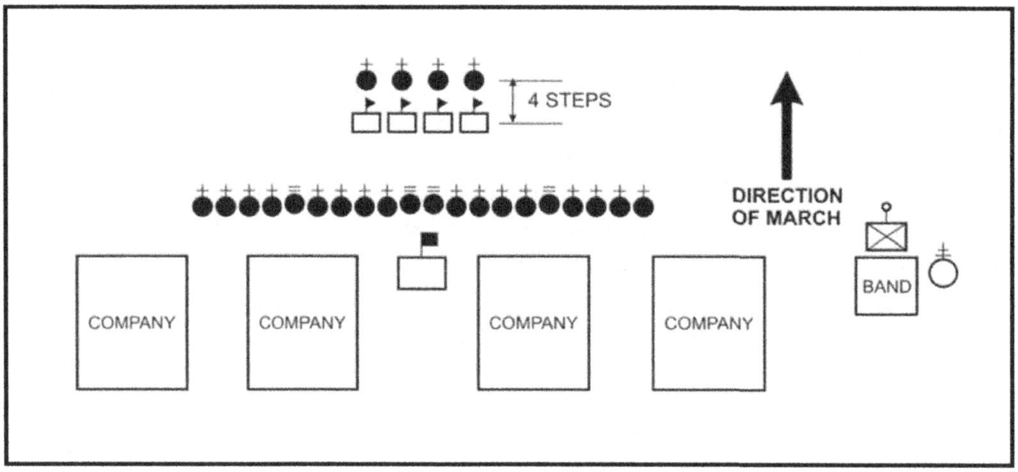

Figure 11-1. Position of key personnel after the command *Officers, Center*, MARCH.

(2) After the officers and guidon bearers have closed and faced to the front, the senior company commander commands ***Forward,*** **MARCH**. The center officer of the leading rank is the guide. The officers and guidon bearers are halted with the leading rank six steps from the battalion commander, and they salute. The senior company commander commands **HALT** and ***Present*** **ARMS**. The battalion commander returns the *Salute.* On the command ***Order,*** **ARMS** by the senior company commander, the officers execute *Order Arms* and the guidon bearers execute *Carry Guidon.*

(3) The battalion commander gives such instruction as he deems necessary and then commands ***Officers, Post*** (pause), **MARCH**.

(a) At the command ***Post,*** all officers and guidon bearers face about.

(b) At the command of execution **MARCH**, the band begins to play and officers and guidon bearers step off.

(4) The senior commander commands ***Officers,*** **HALT.** He halts the leading rank in line, three steps from the line of companies. He then commands ***Post*** (pause), **MARCH.**

(a) At the command ***Post,*** the officers and guidon bearers face outward.

(b) At the command **MARCH**, the officers and guidon bearers step off and return to their posts and come to *Parade Rest.* The music ceases when the last officer has taken his post and come to *Parade Rest.*

(5) During the execution of *Officers Center* and *Officers Post,* except when saluting, all guidon bearers remain at *Carry Guidon.*

i. **Pass in Review.** When all officers have returned to their posts and the band has stopped playing, the battalion commander directs **BRING YOUR UNITS TO ATTENTION.** Unit commanders assume the position of *Attention,* face about, and command ***Company,*** **ATTENTION**. Then they face about. When all units are at *Attention,* the battalion commander directs **PASS IN REVIEW**. The battalion passes in review in the same manner as prescribed for a review except the units render honors when six steps to the left of the commander of troops and staff, and terminate honors when the unit is six steps to the right of the commander of troops and staff. The commander of troops and staff then assumes the role of the reviewing party—at their location on the field.

j. **Conclusion.** When the last element of troops has passed the reviewing stand and executed *Ready Front,* the band ceases playing and goes immediately into an eight-bar drum cadence. On the first beat following drum cadence, the band begins playing (in place) the official version of "The Army Goes Rolling Along," beginning at the introduction. On the second time through letter A (or ninth bar), the drum major has the band march forward and execute a left turn onto the reviewing line while continuing to play until the completion of the chorus. The completion of "The Army Goes Rolling Along" terminates the ceremony. It is appropriate for spectators to rise and stand at the position of *Attention* while "The Army Goes Rolling Along" is being played.

11-4. CEREMONIAL BRIGADE PARADE

The brigade ordinarily is formed in line with battalions in mass formations. The parade is the same as the ceremonial battalion parade with the following exceptions:

- The brigade commander is substituted for battalion commander, and brigade for battalion, in the description.
- In moving across the front of the brigade, the band passes in front of the line of battalion commanders during the *Sound Off.*

a. The battalions execute *Present Arms, Order Arms, Parade Rest,* and come to *Attention* on the command of execution of their respective commanders. Reports are made by the battalion commanders instead of company commanders.

b. At the command of execution **MARCH**, of ***Officers, Center,*** **MARCH**, the battalion commanders, their staffs, and Colors close on the line of battalion commanders and staffs. The company commanders face to the half right or half left in marching and close on a line four steps to the rear of the battalion Color. The guidon bearers face half right or half left in marching and close on a line four steps back of the company commanders. The other officers face half right or half left in marching and close on a line four steps in back of the guidon bearers. The command sergeant major remains in place and assumes command of the battalion.

c. The officers, guidon bearers, and Colors having closed and faced to the front, the senior battalion commander commands ***Forward,*** **MARCH**. The center officer of the leading rank is the guide. On the command of the senior battalion commander, the officers, guidon bearers, and Colors are halted with the leading ranks six steps from the brigade commander. They salute the brigade commander; he returns the *Salute.* The commands ***Present,*** **ARMS** and ***Order,*** **ARMS** are given by the senior battalion commander.

d. The brigade commander commands ***Officers, Post*** (pause), **MARCH.** On the command ***Post,*** the Colors execute *Reverse March.* All others execute *About Face.* On the command of execution **MARCH,** the officers, Colors, and guidon bearers march forward. The senior battalion commander commands ***Officers,*** **HALT** so that the rank of battalion commanders is on line with its original position. The senior battalion commander then commands ***Post*** (pause), **MARCH.** On the command ***Post,*** the battalion commanders and staffs face outward, the battalion Color guard executes wheeling movements, and on the command of execution **MARCH**, the Color guard marches back to its original position. The company commanders, guidon bearers, and other officers face to the half right or half left in marching and move back to their original positions.

11-5. STREET PARADES

For street parades, troops are formed and marched in the most convenient manner. Street parades may include military vehicles. Towed or transported weapons add to the drama of a street parade. Cargo vehicles are included only to increase the size of the display.

a. Practical formations for street parades are:

- Columns of threes and fours.
- Two or more columns of threes and fours abreast.
- Mass formation.

b. The vehicles move in a single column or column of twos, threes, or fours abreast, as the width of the street permits.

c. In order to keep military units in the same cadence, units should not march between two bands.

CHAPTER 12

HONOR GUARDS

In the Continental Army, honor guards were used as a protective measure and as a means of showing the improvement in discipline in the newly formed army. They were probably selected specifically for their size and strength and maybe their proven prowess in combat. Today, the honor guard formation is a special courtesy to visiting dignitaries. Members of the guard are selected for their soldierly appearance and superior discipline.

12-1. BASIC INFORMATION

The basic information herein applies to all honor guards.

a. An honor guard consists of a band, Colors, salute battery (when available and appropriate), and a formation of troops. Honor guards render personal honors to persons of high military or civilian rank or position upon arrival or departure from a military command. Honor guards are not a substitute for those ceremonies appropriate in a review or parade.

b. An honor guard should not be so large as to compromise the exceptional standards required of honor guards. Uniforms will be those prescribed in Army regulations and tables of allowance.

c. The senior commander or his appointed representative is the host and takes part in the ceremony. He is briefed on the sequence of events and advises the person to be honored of actions that will take place.

d. Before the arrival of the person(s) to be honored, the honor guard commander positions the band, formation of troops, and salute battery from right to left in that order. The troops are formed in line of companies or platoons with the Colors centered. When conditions dictate, the salute battery may be located separately but its control and purpose must not be impaired.

12-2. SEQUENCE OF EVENTS AND INDIVIDUAL ACTIONS

The sequence of events for conducting an honor guard are discussed herein.

a. At the approach of the person(s) to be honored, the honor guard commander faces about, commands ***Honor guard,*** **ATTENTION,** and then faces about again. The host welcomes the person(s) to be honored on arrival, escorts him to the position not more than 20 steps in front of and facing the honor guard commander, and takes his position on the guest's left. When the person(s) to be honored has halted in his position, the honor guard commander faces about, commands ***Present,*** **ARMS**, faces about, and salutes.

b. As the honor guard commander executes the *Hand Salute,* the band begins the appropriate honors. When a salute battery is employed, the first round is fired simultaneously with the first note of the music and remaining rounds are fired at three-second intervals. All military personnel in the vicinity of the honor guard formation, except those on security duty, salute during the firing of the *Cannon Salute* and the rendering of honors (AR 600-25).

NOTE: If a foreign dignitary's honors include his national anthem, the sequence of the ceremony is: honors, inspection, and "The Army Goes Rolling Along" concluding the ceremony.

c. On the completion of the honors, the honor guard commander terminates his *Salute*, faces about, and commands ***Order,* ARMS**. He then faces about.

d. At this time, the honoree(s) and host march forward and halt three steps from the honor guard commander. The honor guard commander salutes and reports ***"Sir, the honor guard is prepared for inspection."*** The band begins to play appropriate music upon the *salute* of the guard commander and continues to play until the members of the reviewing party have returned to their posts. The guard commander then guides the honoree(s) and host to the right of the band. The inspecting party passes along the front and rear of the line of troops, to include the salute battery (if used), with the option of omitting the salute battery from the inspection if its distance from the remainder of the honor guard is prohibitive.

e. The honor guard commander takes a position on the right of the person(s) honored and guides him through the inspection. The host takes a position on the right of the honor guard commander.

f. The inspection begins at the right of the band. The inspecting party passes along the front rank of troops. Ranks are not opened; the individual members of the honor guard do not come to *Inspection Arms* or execute *Eyes Right.*

g. The members of the inspecting party render *Hand Salutes* when they pass in front of the Colors.

h. The inspection terminates at the right flank of the band. The honor guard commander commands ***Party,* HALT**. The bandmaster has the band play softly until the reviewing party members begin marching back to their posts. The honor guard commander faces to the half left in marching, takes two steps, halts, and faces about. The host repositions himself to the left of the honoree(s). When the host is in position, the honor guard commander and the honoree(s) exchange *Salutes.* Upon termination of the *Salutes,* the honoree(s) and the host immediately face to the half left in marching and return to their posts. The honor guard commander hesitates momentarily, then faces to the right in marching, and returns to his post.

i. After the inspection is terminated and the honoree(s) and host return to their positions, the honor guard is given ***Present,* ARMS** and the band plays the National Anthem. If the honoree(s) is a foreign dignitary, the national anthem of his country is played first, followed by an approximate three-second drum roll and the playing of "The Star Spangled Banner."

j. When "The Star Spangled Banner" ends, the command ***Order,* ARMS** is given. The honor guard commander faces about after the command for *Order Arms* is given and remains at *Attention* while the band plays one chorus of "The Army Goes Rolling Along," beginning at the introduction. At the completion of "The Army Goes Rolling Along," the honor guard commander salutes and announces ***"Sir, this concludes the ceremony."*** The honor guard remains at *Attention* until the honoree(s) has departed. The band may play incidental background music while the honoree(s) is being introduced to the staff and other guests.

12-3. HONOR CORDON CEREMONIES

In addition to a scheduled honor guard ceremony, honor cordon ceremonies may be used when welcoming or bidding farewell to distinguished visitors. The ceremony consists of a select group of soldiers with characteristics associated with honor guards, designated to honor and provide security at the immediate arrival or departure site.

a. The honor cordon normally consists of two squads, a senior sergeant, a cordon commander, and a host. When deemed appropriate, the commander may elect to use a band, Colors, and distinguishing flags.

(1) ***Arrival.*** Upon arrival of the craft or vehicle, the cordon is marched to a position near the exit (ramp) in two columns, centered on the exit, faced to the center, and dressed with about three steps distance between the ranks. As the honored guest(s) and greeting party depart the immediate vicinity of the exit, and approach within three steps of the host, the cordon commander commands ***Present,*** **ARMS**, and the entire cordon salutes simultaneously. After the honored guest(s) and all greeting party members have cleared the formation, the cordon commander commands ***Order,*** **ARMS**, and the cordon members come to *Order Arms* simultaneously (Figure 12-1, page 12-4).

(2) ***Departure.*** The formation for the honor cordon upon departure is basically the same as that for arrival, except the cordon commander and the senior sergeant position themselves on the flank of the formation away from the departure site. At the approach (about three steps) of the honored guest(s) and host, the cordon commander commands ***Present,*** **ARMS**, and the cordon salutes simultaneously. After the honored guest has entered the conveyance, the cordon commander commands ***Order,*** **ARMS** and remains in place until the conveyance has departed. If the conveyance is an aircraft, the cordon commander commands ***Order,*** **ARMS;** ***Guards,*** **FACE**, and the guards face toward the cordon commander. The honor cordon departs the area.

b. When a band and Colors are a part of the arrival or departure ceremonies, they are positioned as shown in Figure 12-1 (page 12-4). The band plays appropriate music during the ceremony. The organizational Color is dipped in salute when the honored guest(s) approaches to within six steps. After he passes, it is returned to the carry position. As soon as the guest(s) has entered the conveyance during a departure ceremony, the Colors and distinguishing flags depart the area with the honor cordon.

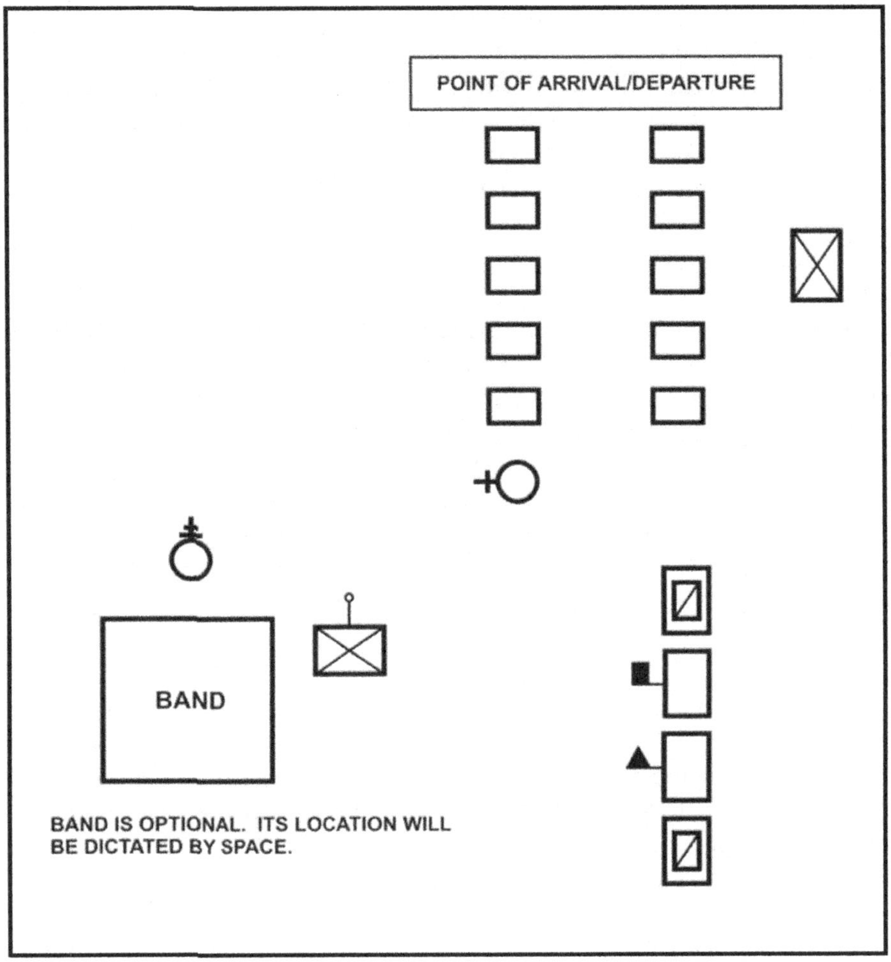

Figure 12-1. Honor cordon formations.

CHAPTER 13
RETREATS AND REVEILLES

Retreat is a ceremony in which the unit honors the U.S. flag when it is lowered in the evening. This ceremony is conducted at the direction of the unit commanders. The installation commander sets the time for sounding retreat. Reveille is a ceremony in which a unit honors the U.S. flag as it is raised in the morning. This ceremony is conducted at the direction of the commander. The installation commander sets the time for sounding reveille. (See Appendix K for more information on raising the flag.)

13-1. RETREAT HISTORY

The term "retreat" is taken from the French word "retraite" and refers to the evening ceremony. The bugle call sounded at retreat was first used in the French army and dates back to the Crusades. Retreat was sounded at sunset to notify sentries to start challenging until sunrise, and to tell the rank and file to go to their quarters. The ceremony remains as a tradition. The old cavalry call "To the Standard," in use from about 1835, has been replaced by the present call of "To the Color." This remains as music honoring the flag as it is lowered in the evening.

13-2. RETREAT SEQUENCE OF EVENTS

The sequence of events for conducting retreat are discussed herein.

a. The unit is formed facing the flag five minutes (if possible) before the sounding of retreat. Four minutes before the sounding of retreat, the adjutant or other appointed officer takes his position centered on and facing the line of troops and commands ***Battalion,*** **ATTENTION** and then ***Parade,*** **REST**.

b. The adjutant faces about and executes *Parade Rest.* On the last note of retreat, the evening gun is fired. The adjutant then comes to *Attention,* faces about, and commands ***Battalion,*** **ATTENTION** and ***Present,*** **ARMS** so that the unit is at *Present Arms* when the first note of "To the Color" or the National Anthem is sounded. The adjutant then faces about and executes *Present Arms*. The adjutant's *Salute* is the signal for the band to begin playing "To the Color."

c. At the last note of "To the Color" or the National Anthem, the adjutant faces about, commands ***Order,*** **ARMS**, and then directs **TAKE CHARGE OF YOUR UNITS**. Unit commanders render the *Hand Salute*. The adjutant returns all *Salutes* with one *Salute*. This terminates the retreat formation.

NOTE: When subordinate units stand retreat, not as a part of a major command, the sequence is the same except the unit commander gives the commands.

13-3. COMMAND RETREAT HISTORY

Command retreat is a ceremony conducted with all members of the command present. Normally, it is conducted by a battalion or larger unit. In the 18th century, command retreat was a daily occurrence, not to honor the flag but as a signal for units to call the roll as a final accounting before reveille the following morning.

13-4. COMMAND RETREAT SEQUENCE OF EVENTS

The sequence of events for conducting command retreat are discussed herein.

a. The unit is formed in line formation facing the flag (if possible). Subunits may be in mass formation. Subunits are brought to *Attention* at the approach of the commander and his staff, or at a predesignated signal.

b. The commander of troops forms his staff in line, takes his position two steps in front of the staff, and marches them to a position centered on and in front of the line of troops. He gives the proper commands to have himself and his staff facing the line of troops when halted.

c. The adjutant, as soon as the staff is halted and without command, moves from his position with the staff to a position midway between the commander of troops and the line of unit commanders. The adjutant commands **REPORT**. All unit commanders salute and report in succession from right to left ***"Sir, company all present or accounted for."*** The adjutant returns each *Salute.* The adjutant then commands ***Present,*** **ARMS**, faces about, salutes, and reports to the commander of troops ***"Sir, all present or accounted for."***

d. The commander of troops returns the *Salute* and directs **POST**. The adjutant moves to his position with the staff. The commander of troops commands ***Order,*** **ARMS**.

e. The left flank member of the staff gives the commands to move the staff from behind the commander of troops to a position two steps in front of the commander of troops. The commander of troops commands ***Parade,*** **REST**. If a band or field music is to be used, the commander comes to *Attention* at the appointed time and commands **SOUND RETREAT**. He then returns to *Parade Rest.*

f. On the last note of "Retreat," the evening gun is fired and the commander of troops commands ***Battalion,*** **ATTENTION** and ***Present,*** **ARMS**. He then executes *About Face* and commands himself and his staff to ***Present,*** **ARMS**. The commander's *Salute* is the signal for the band to begin playing "To the Color" or the National Anthem as the flag is lowered.

g. After the last note of music, the commander of troops commands himself and his staff to ***Order,*** **ARMS**, faces about, commands ***Order,*** **ARMS** and directs **TAKE CHARGE OF YOUR UNITS.** The unit commanders render the *Hand Salute.* The commander of troops returns all *Salutes* with one *Salute.* The commander of troops marches his staff away or dismisses them at this time.

13-5. REVEILLE HISTORY

Reveille was not originally intended as honors to the flag. In 1812, it was a drum call to signify that soldiers should rise for day duty and sentries should leave off night challenging. As time passed, reveille came to denote when the flag was raised in the morning and honors paid to it.

13-6. REVEILLE SEQUENCE OF EVENTS

The sequence of events for conducting reveille are discussed herein.

a. The unit is formed facing the flag five minutes (if possible) before the sounding of reveille. Four minutes before the sounding of reveille, the adjutant or other appointed officer (normally the duty officer) takes his position centered on the line of troops, commands the unit to **ATTENTION,** and commands **REPORT**.

b. All subunits (companies, batteries, or troops) report in succession from right to left, ***"Sir, _______ company all present or accounted for,"*** or ***"Sir, _____ Company, _____ men absent."*** *Salutes* are exchanged with each report.

c. The adjutant commands ***Parade,*** **REST** and then assumes *Parade Rest* himself. If a band is present, about 30 seconds before reveille, the adjutant commands **ATTENTION**, directs **SOUND REVEILLE**, commands ***Present,*** **ARMS**, and then faces about. The adjutant's *Salute* is the signal for the band to sound reveille and to fire the morning gun. When reveille is sounded by a recording, the call **ATTENTION** is sounded about 30 seconds before reveille. This ensures that the adjutant has sufficient time to command the units to ***Present,*** **ARMS** before the first note of reveille.

d. After the last note of reveille has sounded, the adjutant terminates his *Salute,* faces about, commands ***Order,*** **ARMS**, and then directs **TAKE CHARGE OF YOUR UNITS**. The adjutant returns all *Salutes* with one *Salute.* This terminates the ceremony.

NOTE: When a unit conducts the reveille ceremony not as a member of a major command, battalion, or company, the sequence of events remains the same except the unit commander gives all commands.

13-7. COMMAND REVEILLE HISTORY

Command reveille is conducted with all members of the command present. Normally, command reveille is conducted by a battalion or larger unit. Command reveille was conducted as Troop in 1812 and was used to muster the unit or for roll call.

13-8. COMMAND REVEILLE SEQUENCE OF EVENTS

The sequence of events for conducting command reveille are discussed herein.

a. The unit is formed in line formation facing the flag, if possible. Subunits may be in mass formation. Subunits are commanded to *Attention* at the approach of the commander and his staff.

b. The commander of troops takes his post two steps in front of and centered on his staff, commands ***Forward,*** **MARCH**, and marches his staff to a position centered on and in front of the line of troops. The commander of troops gives the proper commands to halt his staff in a position facing the line of troops.

c. As soon as the staff is halted by the commander of troops, the adjutant moves (without command) from his position with the staff to a position midway between the commander of troops and the line of unit commanders. The adjutant commands ***Present,*** **ARMS,** faces the commander of troops, salutes, and reports ***"Sir, the battalion (brigade) is formed."***

d. The commander of troops returns the *Salute* and directs the adjutant to receive the reports. The adjutant faces about and commands ***Order,*** **ARMS** and **REPORT**.

e. Unit commanders salute and report, in succession from right to left, ***"Sir, ______ Company, all present or accounted for."*** The adjutant returns each unit commander's *Salute.* After all unit commanders have reported, the adjutant faces the commander of troops, salutes, and reports ***"Sir, all present or accounted for."*** The commander of troops returns the *Salute* and directs the adjutant to publish the orders.

f. The adjutant faces about without saluting and commands **ATTENTION TO ORDERS.** The adjutant then reads any orders or makes any announcements the commander of troops wishes to publish. The adjutant, on completion of the publishing of orders, faces about and takes his post with the staff without saluting.

g. When the adjutant is in position, the left flank staff member commands ***Right,*** **FACE;** ***Forward,*** **MARCH;** ***Column Left,*** **MARCH;** ***Column Left,*** **MARCH**. When the staff is centered on the commander of troops, he then commands ***Staff,*** **HALT;** ***Left,*** **FACE.** About 30 seconds before the sounding of reveille, the commander of troops commands ***Present,*** **ARMS**, executes *About Face* and commands ***Staff, Present,*** **ARMS**. Reveille is then sounded either by record, field music, or band as the flag is raised.

h. At the first note of "Reveille," the morning gun is fired. After the last note of music, the commander of troops commands ***Staff, Order,*** **ARMS,** executes *About Face,* and commands ***Order,*** **ARMS**; he then directs, **TAKE CHARGE OF YOUR UNITS**. The unit commanders salute. The commander of troops returns all *Salutes* with one *Salute.* He then marches his staff away or dismisses them. This terminates the ceremony.

CHAPTER 14

FUNERALS

The funerals of Soldiers, more than any other ceremony, have followed an old pattern as the living honor the brave dead.

14-1. HISTORY

Funeral services of great magnificence evolved as custom (from what is known about early Christian mourning) in the 6th century. To this day, no religious ceremonies are conducted with more pomp than those intended to commemorate the departed.

a. The first general mourning proclaimed in America was on the death of Benjamin *Franklin in 1791. The second was the death of George Washington in 1799. The deep and widespread grief occasioned by the death of the first President assembled a great number of people for the purpose of paying him a last tribute of respect. On Wednesday, 18 December 1799, attended by military honors and the simplest but grandest ceremonies of religion, his body was deposited in the family vault at Mount Vernon, Virginia.

b. Several military traditions employed today have been brought forward from the past.

(1) Reversed arms, displayed by one opponent on the battlefield, signaled that a truce was requested so that the dead and wounded could be carried off and the dead buried.

(2) Today's customary three volleys fired over a grave probably originated as far back as the Roman Empire. The Roman funeral rites of casting dirt three times on the coffin constituted the "burial." It was customary among the Romans to call the dead three times by name, which ended the funeral ceremony, after which the friends and relatives of the deceased pronounced the word "vale" (farewell) three times as they departed from the tomb. In more recent history, three musket volleys were fired to announce that the burying of the dead was completed and the burial party was ready for battle again.

(3) The custom of using a caisson to carry a coffin most likely had its origins in the 1800s when horse-drawn caissons that pulled artillery pieces also doubled as a conveyance to clear fallen Soldiers from the battlefield.

(4) In the mid to late 1800s a funeral procession of a mounted officer or enlisted man was accompanied by a riderless horse in mourning caparison followed by a hearse. It was also a custom to have the boots of the deceased thrown over the saddle with heels to the front signifying that his march was ended.

*14-2. CLASSES OF MILITARY FUNERALS AND TYPES OF HONORS

Military funerals are divided into two classes: chapel service, followed by movement to the grave or place of local disposition with the prescribed escort; and graveside service only. Burial honors and the composition of funeral escorts are described in Chapter 6, AR 600-25. The types of honors ceremonies that may be performed are described below.

* a. A full military funeral honors normally consist of, or is supported by, a 9-person funeral detail, with the following elements.

- Casualty assistance officer (CAO).
- Officer in charge (OIC) or noncommissioned officer in charge (NCOIC) (appropriate for the rank of the deceased).
- One bugler to play "Taps" (or electronic recording).

- Six active duty pallbearers/firing party (dual function, the pallbearers also serve as the firing party and will render these honors).
- Military clergy (if available and requested).

* b. If resources permit, a larger funeral detail may be provided, which is composed of all the elements of the nine-person funeral detail, and may also include the following.

- Colors.
- Separate firing party (no more than eight, or less than five riflemen).
- Hearse (caisson).
- Honorary pallbearers.
- Personal colors (if appropriate).
- Escort unit(s) (appropriate for the rank of the deceased).

* c. A two-man military funeral honors detail consists of the following elements.

- OIC/NCOIC (appropriate for the rank of the deceased).
- Enlisted soldier.
- One bugler to play "Taps" (or electronic recording).

14-3. INDIVIDUAL RESPONSIBILITIES

*The Casualty Assistance Center (CAC) provides burial honors, for deceased Army personnel, including active duty and retired personnel as well as eligible reserve components and veterans when requested by the family. Active duty Soldiers will receive burial with full military funeral honors, to be provided by a nine-person funeral detail as described in paragraph 14-2 (a). Retirees are entitled to full military funeral honors, resource permitting, but as a minimum, will receive funeral honors consisting of two uniformed Soldiers to fold the flag and present it to the next of kin, and play "Taps." Eligible members of the reserve component and veterans will also receive funeral honors from a two-person detail. Medal of Honor recipients are entitled to full military funeral honors, regardless of status. A live bugler is preferred, however, if none is available, "Taps' may be played on a suitable recording device, but a live bugler is required for all active duty funerals. The family of the deceased (or its representative) may request another clergyman to officiate in lieu of a military chaplain. A civilian clergyman can conduct all religious elements of a military funeral or interment. The desires of the family are given the fullest consideration possible in the selection of elements involved, but the funeral is conducted as prescribed in this manual. For further information, consult AR 600-25, Chapter 6. The responsibilities of the individuals involved in a military funeral are as follows:

a. **Casualty Assistance Office**. The casualty assistance office provides funeral detail requirements and the CAO's name and phone number to the funeral detail NCOIC. It also coordinates bugler commitments.

b. **Funeral Detail NCOIC.** The funeral detail NCOIC—

- Provides the name of the NCOIC and the bugler pick-up time to the casualty assistance office after notification of funeral detail.
- Requests transportation for the funeral detail through the transportation division.
- Coordinates specifics with the funeral home, clergy, and chapel concerned.
- Coordinates the use of a portable CD player for playing "Taps," if needed.
- Ensures all personnel participating in the funeral detail arrive at the designated place in sufficient time to make final coordination.

c. **Transportation Division.** The transportation division provides transportation for funeral details, as required.

d. **Casualty Assistance Officer.** The CAO—

- Coordinates the ceremonial aspects of the funeral.
- Ensures the chaplain receives a flag from the local Post Office or the installation.
- Acts as OIC for the funeral detail and presents the flag to the deceased's next of kin, when required.

e. **Commanding Officer.** The commanding officer or his representative, in coordination with the cemetery superintendent and the funeral director, makes the funeral arrangements and supervises the conduct of the funeral.

14-4. PERSONNEL CONDUCT

Personnel involved with military funerals conduct themselves as described herein.

a. When honorary pallbearers are desired, they are selected by the family of the deceased, or its representative, or by the commanding officer if the family wishes. As a rule, no more than twelve honorary pallbearers should be selected.

b. At a military funeral, persons in military uniform attending in their individual capacity face the casket and execute the *Hand Salute* at the following times: when honors, if any, are sounded; at any time the casket is being moved (the exception being when they themselves are moving); during *Cannon Salutes,* if sounded; during the firing of volleys; and while "Taps" is being played.

(1) Honorary pallbearers in uniform conform to those instructions when not in motion.

* (2) Male military personnel in civilian clothes in the above cases, and during the service at the grave, stand at *Attention,* uncover, and hold the headdress over the left shoulder with the right hand over the heart. If no headdress is worn, the right hand is held over the heart.

(3) Female military personnel in civilian clothes hold the right hand over the heart.

c. During the religious graveside service, all personnel bow their heads at the words "Let us pray." All mourners at graveside, except the active pallbearers, follow the example of the officiating chaplain. If he uncovers, they uncover; if he remains covered, they remain covered. When the officiating chaplain wears a biretta (clerical headpiece) during the graveside service, all personnel, as indicated above, uncover. When the officiating chaplain wears a yarmulke (Jewish skull cap), all personnel remain covered.

d. The remains of a member of the armed forces who died while on active duty, may be consigned directly to a national cemetery from a military installation. In such cases, the cemetery superintendent will, regardless of time of arrival, if not otherwise provided for, engage a funeral director to receive the remains at the common carrier terminal, hold the remains at his establishment until the date of the funeral, if necessary, and deliver the remains to the cemetery. The superintendent will not authorize a funeral director to render any other service incident to the interment.

* e. The word "chapel" is interpreted to include the church, home, or other place where services are held, other than the service at the grave. The word "casket" is interpreted to include the receptacle containing the cremated remains of the deceased.

*14-5. FUNERAL WITH CHAPEL SERVICE (FULL MILITARY HONORS-NINE PERSON FUNERAL HONORS DETAIL)

Use the following procedures to conduct a funeral in a chapel with full military honors.

* a. At the chapel, the funeral detail forms as shown in Figure 14-1. The NCOIC has all participants at the position of *Parade Rest*. The NCOIC and the pallbearers will be on line at normal intervals facing the chapel and close to the designated arrival point of the conveyance. The NCOIC positions himself at the end of the pallbearers so that the conveyance passes him first as it approaches. If resources permit and there is a separate firing party, they will form two ranks facing each other and form an aisle from the conveyance to the entrance of the chapel.

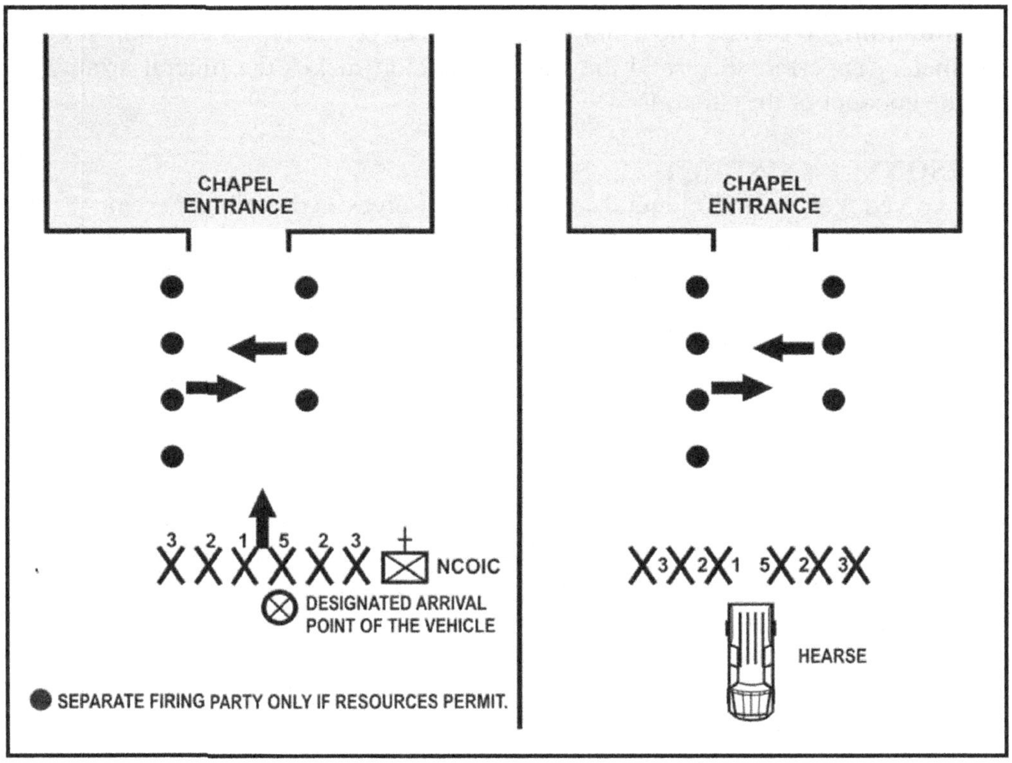

*** Figure 14-1. Funeral detail formation.**

b. Members of the immediate family, relatives, friends of the deceased, and the CAO will be seated in the chapel before the conveyance arrives and the casket is taken into the chapel. Members of the immediate family and relatives occupy pews (seats) to the right (front) of the chapel.

* c. As the conveyance comes into view, the NCOIC will command the pallbearers to **ATTENTION** and render a solo hand salute as the conveyance approaches. Once the conveyance stops in front of the chapel, the NCOIC will automatically **Order Arms**. If a separate firing party exists the following actions will be taken. The NCOIC commands ***Escort,*** **ATTENTION;** ***Pallbearers, Center,*** **FACE.** On the command ***Center,*** **FACE**, the pallbearers face the designated arrival point of the conveyance (Figure 14-1, page 14-4). As the conveyance approaches, the NCOIC commands ***Present,*** **ARMS** and salutes to honor the National Colors draped over the casket and commands ***Order,*** **ARMS** after the conveyance halts.

d. If necessary, the NCOIC repositions the pallbearers at the rear of the conveyance.

* e. After the funeral director opens the doors of the hearse, the NCOIC and the firing party, if available, *Present Arms*. The firing party and the NCOIC *Present Arms* until the casket enters

the chapel. The senior pallbearer, designated position 5, and the pallbearer in position 1 grasp the handles at the head of the casket. (The union of the flag is draped over this end.) They walk backwards, pulling the casket from the conveyance, allowing the pallbearers in positions 2 and 3 to grasp handles on the casket. The pallbearers handle the remains in a dignified, reverent, and military manner, ensuring the casket is carried level and feet first at all times (Figure 14-2).

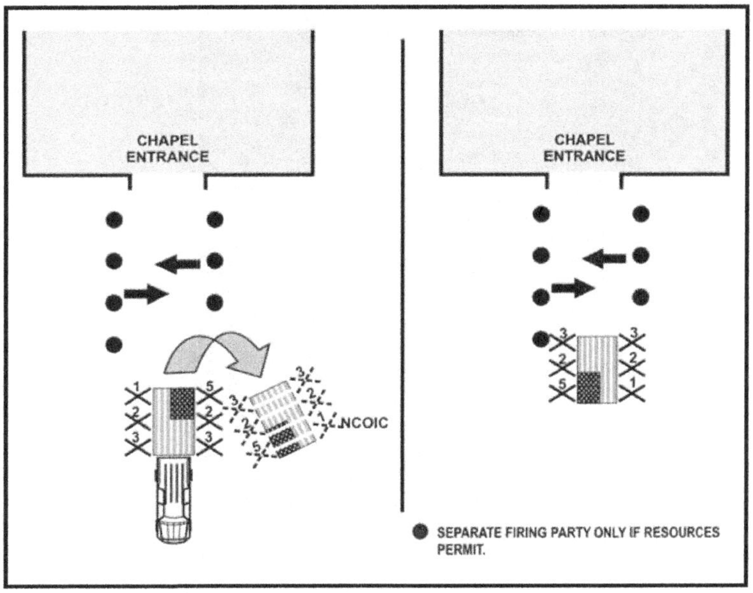

***Figure 14-2. Carrying the casket inside.**

* f. For funerals where there is a separate firing party, once the casket is borne between the firing party members, and taken into the chapel, the NCOIC commands ***Order*, ARMS.** The firing party departs under the control of the firing party commander and travels to the gravesite. Once at the gravesite, the firing party makes preparations for the gravesite ceremony. The bugler, if not already at the gravesite, travels with the firing party.

g. Having entered the chapel, the pallbearers carry the casket to the front of the church. If a church truck is available, the casket is placed on the truck at the entrance of the chapel and pushed to the front by the senior pallbearer and one other. The pallbearers then take seats, as directed by the chaplain, until the conclusion of the chapel service.

* h. Displaying the Flag on the Casket. For information on how to display the US Flag on the casket, either closed or half-couch, consult DA Pamphlet 638-2, Appendix E.

i. After the service, the pallbearers either carry the casket or push it on a church truck from the front of the chapel to the exit. The casket is placed directly into the conveyance with the senior and number 1 pallbearers being the last to release their casket handles. The funeral director secures the doors of the conveyance.

* j. The pallbearers board their transportation and travel to the interment site to prepare for the graveside ceremony. The funeral party travels in the following order (Figure 14-3, page 14-7):

- Clergy.
- Conveyance with casket.

- Active pallbearers.
- Personal flag (if appropriate).
- Family and CAO.
- Friends.

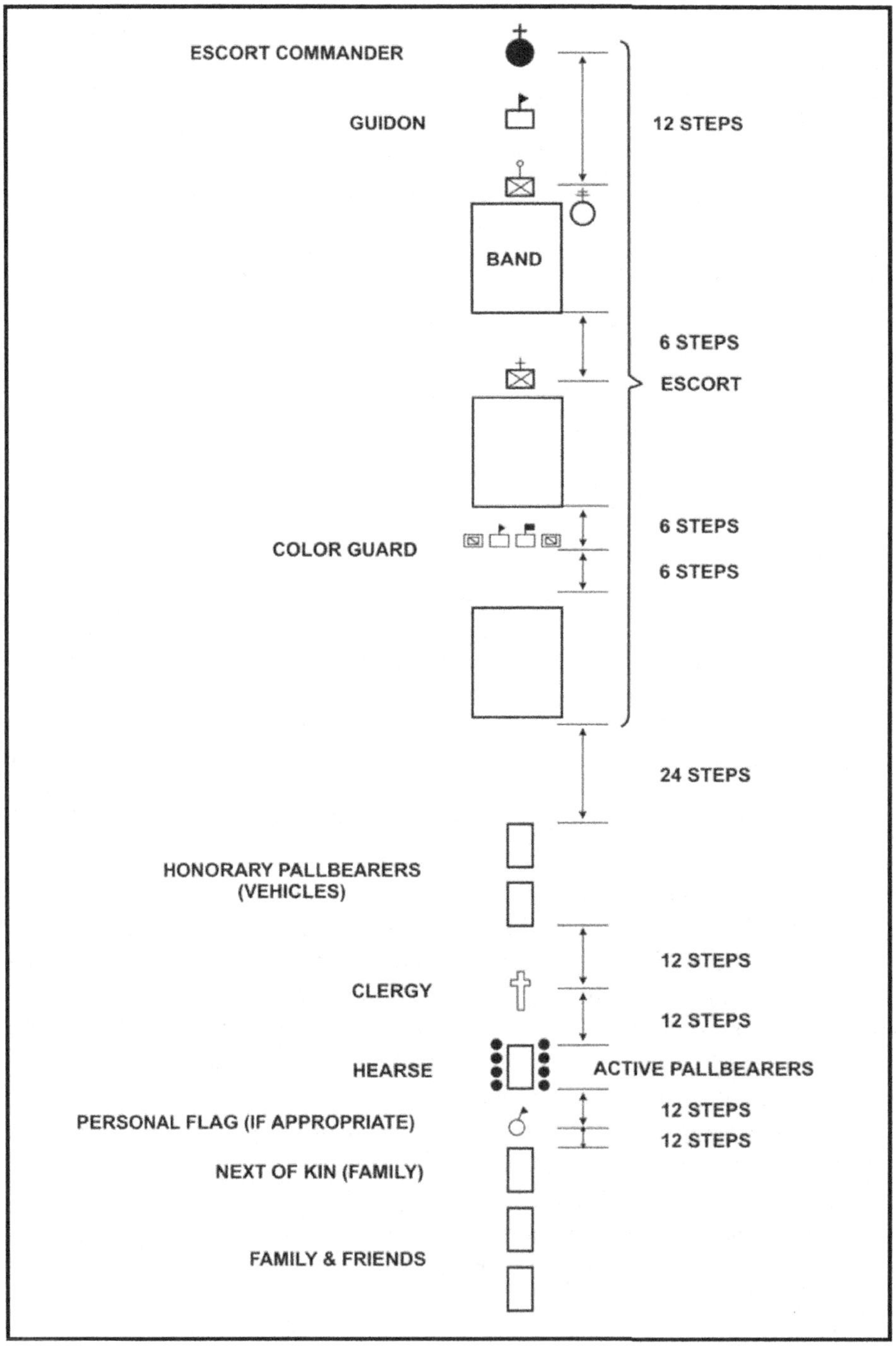

Figure 14-3. Funeral procession.

k. After the procession is formed, it travels directly to the gravesite. Upon arrival, the CAO positions himself between the chaplain and the head of the gravesite. The pallbearers form and remove the casket from the conveyance the same as previously outlined (Figure 14-4).

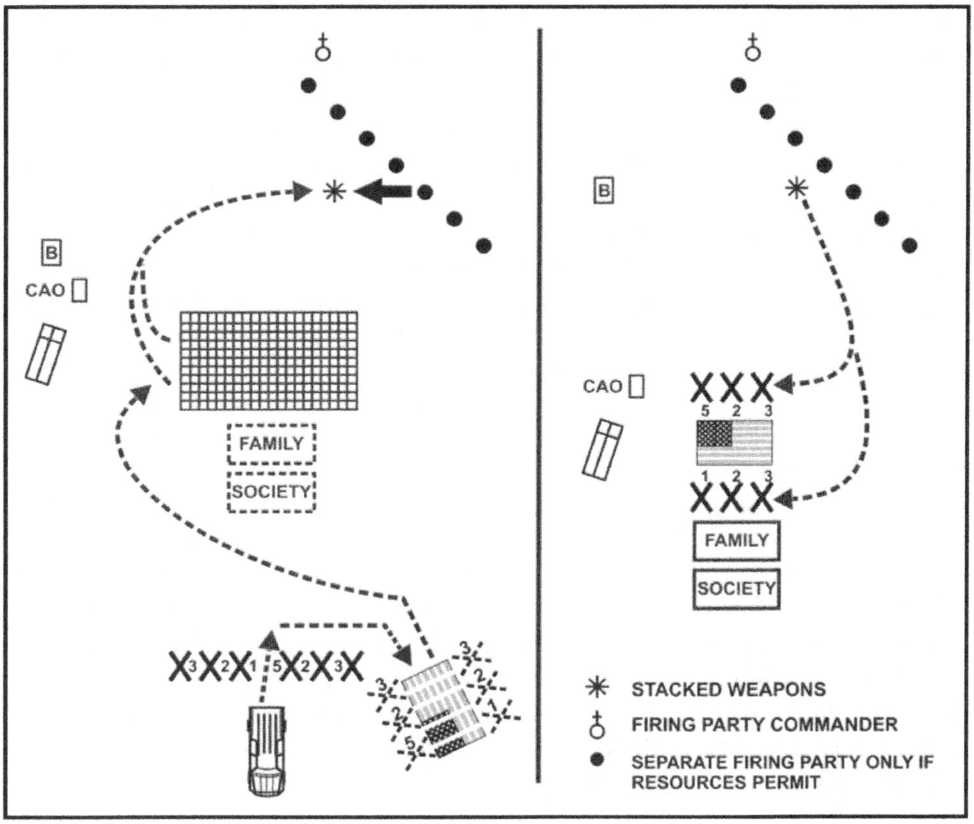

***Figure 14-4. Graveside formation.**

* l. Once the casket is removed from the conveyance, the NCOIC commands the firing party (if resources permit a separate firing party) and bugler to *Present Arms*.

* m. The pallbearers carry the casket, feet first, to the grave. Upon reaching the grave, the casket is placed on the lowering device. The pallbearers then execute the appropriate facing movement and march off in two ranks toward the designated firing party location. While marching, the pallbearers merge into single file in order to form one rank, 45 degrees off the foot of the casket. The firing party commander is positioned on the opposite flank or centered to the rear of the formation (Figure 14-4). At that time, the firing party commander will command the firing party to unstack their weapons, which have been pre-positioned under guard at the gravesite, and then to stand at *"Parade Rest"* during the gravesite service.

* n. When the casket is placed over the grave, and the pallbearers march from the casket to become the firing party, the NCOIC terminates his salute and moves from his place at the head of the casket in order to permit the chaplain to conduct the graveside service. He should move to a location where he still faces the family, but does not interfere with the service. Once in position, he assumes the position of *Parade Rest* until the service is completed.

* o. After *Parade Rest* has been commanded, the chaplain conducts the graveside service. At the conclusion of the benediction, the NCOIC returns to his position at the head of the casket, renders a hand salute, which also cues the firing party commander to commence the firing party sequence, as outlined in paragraph 14-17. The CAO also executes *Present Arms*. The firing party fires three volleys of blank cartridges, assumes the position of *Present Arms* at the command of the firing party commander, and remains in this position until the conclusion of "Taps." The bugler, positioned near the firing party and in view of the next of kin, sounds "Taps" immediately following the firing party assuming *Present Arms*.

* p. At the conclusion of "Taps," the firing party comes to *Order Arms* at the command of the firing party commander, and restacks their weapons in a ceremonial manner. After the stacking of weapons is completed, the firing party forms into two ranks and marches in the most direct route back to the lowering device platform so they can perform the ceremonial folding of the interment flag (Figure 14-4). The CAO terminates his *Salute*.

* q. The pallbearers raise the flag from the casket and hold it in a horizontal position waist high and complete the folding sequence without letting the flag touch the casket. As the flag is folded, it is passed to the senior pallbearer at the head of the casket, who makes the final tuck. (See Appendix K for detailed information on folding the flag.)

* r. After the flag is folded, the senior pallbearer executes a *Right Face* and places the flag at chest level into the hands of the CAO. The CAO salutes the flag for three seconds before accepting it from the senior pallbearer. The senior pallbearer salutes the flag for three seconds after presenting it to the CAO. The CAO then moves by the most direct route to the next of kin who is to receive the flag. Upon presentation, the CAO renders appropriate remarks such as, "Sir/Ma'am, this flag is presented on behalf of a grateful nation as an expression of appreciation for the honorable and faithful service rendered by your loved one." After the flag is presented, the CAO returns to his original position.

* s. After the presentation is completed, the NCOIC marches the pallbearers and the bugler away from the gravesite and toward the stacked weapons. At the first *Halt,* the rifles of the firing party are then cleared and inspected, which concludes the ceremony.

14-6. TWO-MAN HONOR DETAIL

A two-man honor detail provides graveside honors by the playing of "Taps" and the flag folding and presentation to the appropriate family member. Use the following procedures to conduct a military funeral with a two-man honor detail.

* a. Once the Army CAC is alerted, it arranges for the two-man military honor detail to arrive at the interment site at the appropriate time to provide graveside honors.

* (1) The leader of the detail has many responsibilities to include contacting the funeral director to confirm the date, time, and location of the interment service. The leader ensures that the funeral director has obtained a flag for the ceremony. The detail leader will bring a backup flag to the ceremony in case it is needed.

(2) The leader confirms and coordinates participation of the second member of the detail.

(3) When all coordination is completed, the final preinterment activity is to train and rehearse the detail. A mandatory training item is to carefully watch a video demonstration tape provided by DOD to each installation.

(4) On the day of the interment ceremony, the detail leader confirms arrangements with the funeral director and coordinates necessary cues at the interment site.

* b. The rendition of "Taps" may be by bugler or by electronic device.

* (1) The CAC actively searches for a bugler. (Military or civilian may be used.) Bugler support may be from an Army band (Active or Reserve component), contracted, or voluntary.

(2) If a bugler is not available, the CAC uses the high-quality recording of the U. S. Army band bugler provided by OSD on compact disk. Many national and private cemeteries have sound systems that play "Taps" at the interment site. However, CACs cannot assume availability of such systems and must have a sufficient number of high-quality, portable CD players to provide their own sound system at funerals. (A portable CD player that can be easily heard by all attendees at the interment ceremony is recommended.) Before departing for a funeral, the detail leader must determine if a sound system is available or if the CAC must provide a sound system to the honors detail.

c. The detail arrives at the interment site early and conducts a reconnaissance and rehearsal. Part of the reconnaissance is the selection of a location for the bugler or CD player that will sound "Taps." The detail leader sets up and tests the CD player, ensuring the unit and its remote controls are working properly and that it is out of sight of the family.

(1) When everything is prepared, the detail leader positions the detail in their designated place before the arrival of the funeral cortege. The detail leader positions himself near the recording device; the other members(s) will be positioned near the foot of the grave.

(2) The leader brings the team to *Attention* and *Present Arms* as the remains are carried to the gravesite by civilian pallbearers. He commands ***Order,*** **ARMS** when the casket is placed on the lowering device.

(3) At the conclusion of the committal service, the detail leader sounds "Taps" electronically or directs the bugler to sound "Taps." Installations must ensure that honor detail training directs that the recording device be positioned out of sight of the family and be played in a dignified manner as shown in the training video from DOD.

(4) Although the CD player should be out of sight, activating the "play" button should be performed with precision and distinction by bending over, activating the recorder, and then stepping back one step and assuming the *Position of Attention.*

(5) Each detail member will *Present Arms* during "Taps" and will execute *Order Arms* at its completion. At the conclusion of "Taps," the detail leader ensures the recording device is turned off and then proceeds in a dignified and military manner to the head of the casket.

* d. For flag folding, upon conclusion of "Taps," the representative and his assistant move closer to the casket. When the flag is secured and raised, the detail takes three steps away from the mourners and folds the flag. (See Appendix K for detailed information on folding the flag.) When the flag is properly folded, the detail leader salutes the flag for three seconds. The assistant hands the flag to the detail leader, salutes the flag for three seconds, and posts to a position next to the side or rear of the family. After the assistant departs, the detail leader presents the flag to the next of kin using the following wording: "Sir/Ma'am, this flag is presented on behalf of a grateful nation as an expression of appreciation for the honorable and faithful service rendered by your loved one." After presenting the flag, the detail leader offers condolences.

* e. Not all funerals will be authorized the human resources as outlined in this sequence of events; therefore the CAO and NCOIC will extract those portions of the sequence that apply to their funeral detail contingent.

f. Additions to an element of the funeral detail not specifically addressed in this sequence of events are not authorized. Requests for exceptions to policy will be directed to TRADOC.

***NOTE:** If a military chaplain is present, he/she presents the flag to the next of kin.

14-7. GRAVESIDE SERVICE

*For a funeral without chapel service, all elements of a military funeral are present and used as previously described. However, if troops are not conveniently available, or if the family wishes to eliminate other elements, the following are used (Figure 14-5):

- *Military clergy (if available and requested).
- Officer in charge or noncommissioned officer in charge, appropriate to the grade of the deceased (AR 600-25).
- Active pallbearers/firing party.
- Separate firing party (if resources permit).
- Bugler.
- Personal Color bearer (if appropriate).

These elements are in position at the graveside before the arrival of the remains.

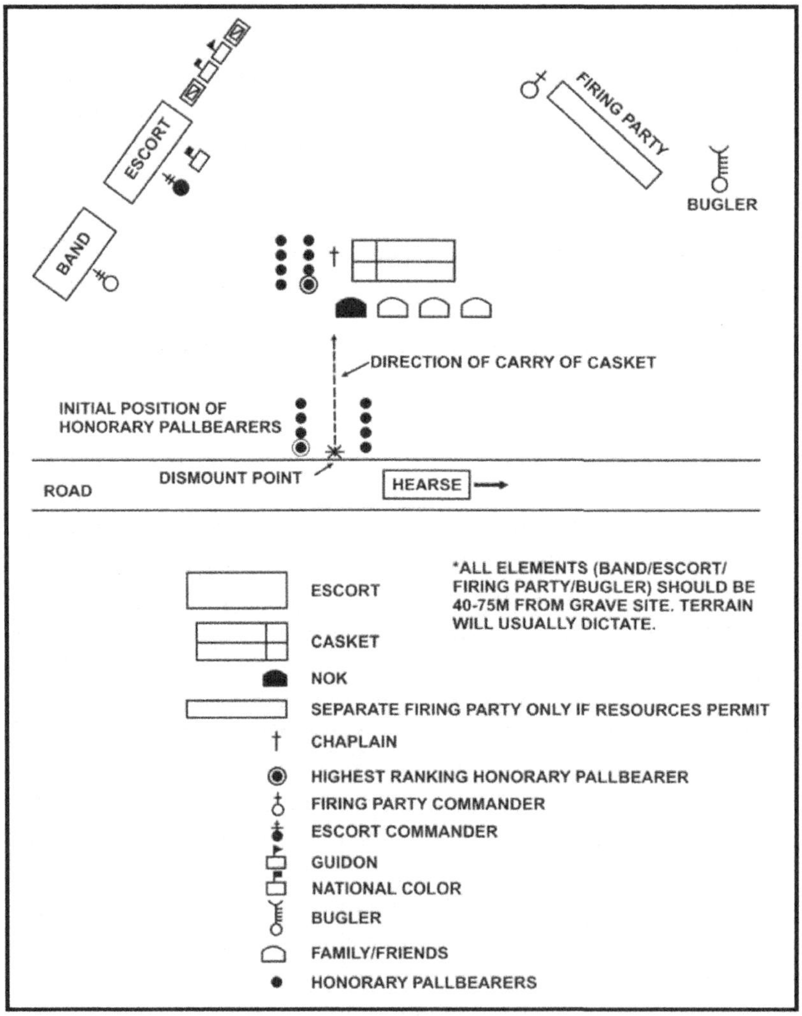

***Figure 14-5. Graveside service.**

14-8. CREMATED REMAINS

When the remains are cremated and the ashes interred with military honors, the previously stated provisions, with necessary modifications, will govern.

* a. For all phases of the funeral, where the cremated remains are carried by hand, one pallbearer is detailed to carry the receptacle (casket) containing the ashes and another is detailed to carry the flag, folded into the shape of a cocked hat. The pallbearer carrying the flag is always positioned to the right of the remains (Figure 14-6, page 14-13). When the receptacle is carried from the hearse into the chapel and from the chapel to the hearse, these two pallbearers are the only participants in the ceremony. During the procession to the gravesite, the receptacle and flag are carried by the two pallbearers followed by four additional pallbearers. When the receptacle has been placed on the gravesite, all six pallbearers unfold the flag and hold it over the grave. (Honors are the same as a flag- draped casket.)

Figure 14-6. Pallbearers for cremated remains.

b. When the receptacle and flag are placed before the chancel of the chapel or transported to gravesite by vehicle, the receptacle and folded flag are placed side by side. If the pallbearers walk to the gravesite, the two bearers who carried the receptacle and the flag join the other four pallbearers already pre-positioned on either side of the hearse.

c. When no hearse is used, suitable transportation is provided for the receptacle and flag bearers, and the other pallbearers.

d. When the remains are moved to a crematory and the ashes are to be interred with military honors at a later time, the ceremony consists only of the escort to the crematory. All personnel salute as the remains are carried into the crematory. The firing of volleys and the sounding of "Taps" are omitted. When the funeral ceremony is held at the crematory, and when no further honors are anticipated, the volleys are fired and "Taps" is sounded at the discretion of the commanding officer.

NOTE: In this situation, the flag is carried left hand over right hand with the point forward.

14-9. CEREMONY BEFORE SHIPMENT OF REMAINS

When the remains of a deceased soldier are moved to a railway station or other point for shipment to another place for interment or final disposition, funeral services are modified as necessary. When no further military honors are anticipated at the place of interment or final disposition, the volleys are fired and "Taps" sounded at the discretion of the commanding officer. When military honors are anticipated at the place of final disposition, the volleys and "Taps" are omitted.

14-10. CANNON SALUTE

When the funeral of a general officer on the active or retired list, who was entitled to a *Cannon Salute,* takes place at or near a military installation, guns equal to the number to which the officer was entitled (AR 600-25) may be fired at noon on the day of the funeral. The military installation mentioned in general orders will fire the prescribed *Salutes.* Immediately preceding the benediction, a *Cannon Salute* corresponding to the grade of the deceased (AR 600-25) is fired at five-second intervals. Following the benediction, three volleys of musketry are fired.

14-11. PARTICIPATION OF AVIATION

When aviation participates in a military funeral, it is timed so that the aircraft appear over the procession.

*14-12. PARTICIPATION OF RESERVE COMPONENTS

The Reserve Component (RC), along with the active Army, are required to participate in funeral details. The Army National Guard (ARNG) and U. S. Army Reserve (USAR) have a single point of contact (POC) in each ARNG state area command (STARC) or USAR Regional Support Command (RSC) to which a request for assistance can be made. When the active Army is unable to support the request, or it is more prudent for the RC unit to provide honors, the CAC contacts the RC POC at either the STARC or RSC for military funeral honors support. If the RC POC does not respond to the request for support within two hours, the CAC should again contact the RC POC. When the RC is unable to support the request for assistance, the CAC is responsible for providing the honors. The casualty and memorial affairs operations center, PERSCOM will provide a list of RC POCs to the CACs. CACs should establish memorandums of agreement with RC POCs and other military organizations within their area of responsibility specifying requirements and responsibilities.

14-13. PARTICIPATION OF FRATERNAL OR PATRIOTIC ORGANIZATIONS

The family or representative of the deceased may request fraternal or patriotic organizations, of which the deceased was a member, to take part in the funeral service. With immediate family approval fraternal or patriotic organizations may conduct graveside service at the conclusion of the military portion of the ceremony, signified by the flag presentation to the next of kin and escort departure from the cemetery.

14-14. DUTIES OF THE CHAPLAIN

The chaplain takes his position in front of the chapel before the arrival of the remains. He precedes the casket when it is carried from the hearse into the chapel and from the chapel to the hearse. While the remains are being placed in the hearse, he stands at the rear and to the side facing the hearse. When he is wearing vestments, he may, at his discretion, proceed from the chancel to the sacristy (vestry) at the conclusion of the chapel service and divest, joining the procession before it moves from the chapel. He then precedes the hearse to the graveside and precedes the casket to the grave.

14-15. PRELIMINARY ARRANGEMENTS

The officer in charge of a military funeral, the commander of the escort, the funeral director, and the superintendent of the cemetery or his representative visit the places involved and make careful arrangements before the time set for the funeral. They determine the positions at the grave for the various elements of the funeral and make arrangements for traffic control.

14-16. FLORAL TRIBUTES

In the absence of the chaplain, the chaplain's assistant helps the funeral director in arranging all floral tributes in the chapel. The commanding officer or his representative coordinates the necessary transportation with the funeral director for prompt transfer of floral tributes from the chapel to the gravesite. The vehicle bearing the floral tributes is loaded promptly at the conclusion of the chapel service. It precedes the funeral procession, moving as rapidly as practicable to the site of the grave. The funeral procession does not move from the chapel until the vehicle carrying the floral tributes has cleared the escort. The funeral director or the cemetery representative is responsible for removing cards and making a record that gives a brief description of the floral piece pertaining to each card. After completion of the funeral services, the cards and records are turned over to a member of the family of the deceased.

14-17. RULES FOR CEREMONIAL FIRING

For ceremonial firing, the firing party consists of not more than eight riflemen and not less than five with one noncommissioned officer in charge (Figure 14-7, page 14-16). The firing party is normally pre-positioned at the gravesite and facing in the direction that allows it to fire directly over the grave. However, care should be taken to ensure that rifles are fired at a 45-degree angle from the horizontal.

a. To load:

(1) Magazines or clips are loaded with three rounds and blank adapters are attached before forming the firing party.

(2) At the conclusion of the religious services or on the escort commander's command, the noncommissioned officer in charge commands ***With blank ammunition,*** **LOAD**. At the command **LOAD**, each rifleman executes *Port Arms*, faces to the half right, and moves his right

foot 10 inches to the right to a position that gives him a firm, steady stance. He then chambers a round, places the weapon in the safe position, and resumes *Port Arms*.

b. To fire by volley:

(1) When the riflemen have completed the movements and the weapons are locked, the commands are ***Ready, Aim,*** **FIRE**. At the command ***Ready,*** each rifleman moves the safety to the fire position. On the command ***Aim,*** the rifle is shouldered with both hands with the muzzle to the front at an angle of 45 degrees from the horizontal. On the command of execution **FIRE**, the trigger is squeezed quickly, and the weapon is immediately returned to *Port Arms*.

(2) To continue the firing with weapons that function automatically (blank adapter), the commands ***Aim*** and **FIRE** are given and executed as previously prescribed. To continue the firing with weapons that must be manually operated to chamber another round (without blank adapters), the commands ***Ready, Aim,*** **FIRE** are again given. On, the command ***Ready,*** each rifleman manually chambers the next round. The commands ***Aim*** and **FIRE** are then given and executed as previously prescribed.

* (3) When the third round has been fired and the riflemen have resumed *Port Arms*, the noncommissioned officer in charge commands **CEASE FIRING**. The riflemen immediately place the weapon on safe, assume the *Position of Attention* (at *Port Arms),* and face to half left. From this position, the firing party is commanded to *Present Arms* before the playing of "Taps." After "Taps," they are commanded to *Order Arms*. The noncommissioned officer in charge executes a *Right (Left) Face* and remains at *Attention* until the flag has been folded and saluted by the officer in charge or noncommissioned officer in charge of the funeral detail. At this time, the firing party noncommissioned officer in charge executes a *Right (Left) Face* and commands ***Right (Left),*** **FACE;** ***Port,*** **ARMS;** and ***Forward,*** **MARCH**. At the first halt, the rifles of the firing party are cleared and inspected.

NOTE: THE NCOIC MAY BE POSITIONED ON THE OPPOSITE FLANK OR TO THE REAR OF THE FORMATION.

Figure 14-7. Position of fire, ceremonial firing.

CHAPTER 15
COLORS

From the earliest times, warriors used a banner or other symbol to identify specific units and to serve as a rallying point for troops. In medieval days, the standard or banner was used to signal a general assault, which was generated by a cry of "Advance your banners." Until comparatively recent years, the flags that identified nations usually were based on the personal or family heraldry of the reigning monarch or ruling nobleman. As autocracies faded or disappeared, dynastic colors were no longer popular and national flags, as thought of today, came into being. These national flags, such as the Union Jack of Great Britain, the Tricolor of France, and the Stars and Stripes, are relatively new to history.

15-1. HISTORY

Flags are almost as old as civilization itself. Imperial Egypt as well as the armies of Babylon, Chaldea, and Assyria followed the colors of their kings. The Old Testament frequently mentions banners and standards. Many flags of different designs were present in parts of the American colonies before the Revolution. When the struggle for independence united the colonies, the colonists wanted a single flag to represent the new nation. The first flag borne by the Army as a representative of the 13 colonies was the Grand Union flag. It was raised over the Continental Army at Cambridge, Massachusetts, on 2 January 1776. That flag had the familiar 13 stripes (red and white) of the present flag, but the blue square contained the Crosses of St. George and St. Andrew from the British flag.

a. The Stars and Stripes was born on 14 June 1777, two years to the day after the birth of the Army. On that date, Congress resolved that the flag of the United States be 13 stripes, alternate red and white, and that the union be 13 stars, white in a blue field, representing a new constellation. The arrangement of the stars on the blue field was not specified.

(1) According to some historians, the Stars and Stripes was first raised over Fort Stanwix, New York, on 3 August 1777. In that Army version of the flag, the stars were arranged in a circle. (The Navy version had the stars arranged to form crosses similar to the British flag.)

(2) When Vermont and Kentucky joined the Union, the flag was modified so that there were 15 stars and 15 stripes. It was that flag, flying triumphantly over Fort McHenry, Maryland, on 13 and 14 September 1814, which inspired Francis Scott Key to compose the verses of "The Star Spangled Banner." That flag was the national banner from 1795 until 1818. Thus, when it was raised over Tripoli by the Marines in 1805, it was the first United States flag to be hoisted over conquered territory in the Old World. Later, it was flown by General Andrew Jackson at the Battle of New Orleans.

(3) Realizing that adding a stripe for each new state would soon spoil the appearance of the flag, Congress passed a law in 1818 fixing the number of stripes at 13 and providing for the addition of a star in the Blue Union for each new state. The star is to be

added and the new flag to become official on the Fourth of July following the admission of the new state to the Union.

(4) It was not until shortly before the Civil War that the Stars and Stripes actually became the National Color.

b. Before the Civil War, in lieu of a National Color, the US soldiers carried a blue silk color on which was embroidered the arms of the United States, and an American eagle bearing a shield on its breast, and in its talons an olive branch and arrows, signifying peace and war. After the National Color was authorized, the organizational color with the eagle became the regimental color. Because of the high casualty rate among the members of the Color party, plus the advent of modern weapons, the time-honored practice of carrying the Colors in battle was discontinued. Today, the Colors, with battle streamers attached, join their unit in formations during ceremonies to signify their presence during past battles.

15-2. THE COLOR AND COLORS

The National and organizational flags carried by Color-bearing units are called the National Color and the organizational Color. When used singularly, the term Color implies the National Color. The plural term Colors means the national and positional or organizational Colors. By regulation, the organizational Color is not authorized a *Salute;* however, in the past some organizations have observed the custom of saluting the organizational Color.

a. In garrison, the Colors are normally kept at the office or headquarters of the commanding officer. They are escorted to and from the headquarters or office by the Color guard. In the field, the Colors are normally displayed from reveille to retreat in front of the commanding officer's tent or command post. During inclement weather, they are cased and placed in the commanding officer's office, headquarters, or tent.

b. Individuals or units passing or being passed by uncased Colors out of doors render honors. Individuals, not part of a formation, salute six steps distance from the Colors and hold the *Salute* until they have passed six steps beyond the Colors. The individual in charge of a formation calls the formation to **ATTENTION** and ***Present,* ARMS**.

c. The Colors may be carried in any formation in which two or more companies, honor guards, or representative elements of a command participate.

d. Traditionally, the command sergeant major is responsible for the safeguarding, care, and display of the organizational Color. He is also responsible for the selection, training, and performance of the Color bearers and Color guards.

e. The regulation for individual flags and distinguishing flags for general officers and other dignitaries is the same as that for organizational Colors (AR 840-10).

f. During a review, parade, or honor guard ceremony, ordinarily only one National Color is present. The National Color is given the honor position and is carried on the marching right of positional and organizational Colors. The United States Army flag or the Army field flag (when authorized) is carried to the immediate left of the National Color. The organizational Color of the senior headquarters sponsoring the ceremony is carried to the left of the Army flag or field flag, if present. The Colors belonging to the headquarters conducting the ceremonies are positioned on line with and centered on the command. Subordinate Color-bearing organizations ordinarily carry only their organizational Colors (four steps to the rear of their staff).

g. When the local commander wishes, or considers it more appropriate, he may authorize subordinate Color-bearing organizations to carry their National Color with their organizational Colors behind their staffs. Consideration must be given as to the number of times the spectators will be required to stand during the pass in review. Also, consideration must be given to the frequency of *Salutes* required by the reviewing officer.

(1) Another method in which subordinate Color-bearing units may carry their Colors is to have them participate in a massed Colors formation. Because of many complications caused by massed Colors, units should prescribe a local SOP governing the desired procedures.

(2) Examples of complications include: organizational Colors lose their identity, order of precedence, resizing of Color guards and bearers, uniformity when the first rank consists of five men rather than four, and presenting a good appearance when returning to the formation after being brought forward while keeping the National Color on the marching right.

h. When Colors are participating in a ceremony they will be received and dismissed as described in paragraphs 15-5 and 15-7.

15-3. SALUTES

The National Color renders no *Salute* (dip). An exception to this rule is followed by naval vessels when, upon receiving a *Salute* of this type from a vessel registered by a nation formally recognized by the United States, the compliment must be returned.

a. The organizational Color salutes (dips) in all military ceremonies while the National Anthem, "To the Color," or a foreign national anthem is being played, and when rendering honors to the organizational commander or an individual of higher grade including foreign dignitaries of higher grade, but in no other case. The United States Army flag is considered to be an organizational Color and, as such, is also dipped while the National Anthem, "To the Color," or a foreign national anthem is being played, and when rendering honors to the Chief of Staff of the United States Army, his direct representative, or an individual of equivalent or higher grade, but in no other case.

b. When marching, organizational Colors salute when six steps from the person entitled to the *Salute.* They are returned to the *Carry* position when six steps beyond the person.

15-4. COLOR GUARD

The Color guard consists of two (three) sergeants and two specialists or privates. It is an honor to be selected as a member of the Color guard. The senior (Color) sergeant carries the National Color and commands the Color guard. He gives the necessary commands for the movements and for rendering honors.

a. When battalions or brigades carry their organizational Colors in a ceremony as part of a larger command, the battalion or brigade Color is carried four steps to the rear of the staff. A sergeant acts as Color bearer and two experienced specialists or privates, selected by the battalion or brigade command sergeant major, act as members of the Color guard.

b. The Color guard is formed and *Marched* in one rank at *Close Interval,* the bearers in the center. They do not execute *Rear March* or *About Face.* The Color guard marches at *Right Shoulder Arms* and executes *Facing* movements by wheeling to the right or left. The command for a *Facing* movement is ***Right (Left) Wheel,*** **MARCH**. To execute a *Wheeling* movement, the guard nearest the direction of turn serves as the pivot point and executes the movement by marching in place and simultaneously turning in the new direction. Other members shorten their steps and turn in an arc keeping abreast of each other to maintain alignment. When the movement has been completed, each member automatically marches in place until the command **HALT** or ***Forward,*** **MARCH** is given.

c. When passing in review, the Color guard executes *Eyes Right* at the prescribed saluting distance on the command of the Color sergeant. The commands are ***Eyes,*** **RIGHT** and ***Ready,*** **FRONT**. The organizational color salutes at the command **RIGHT**, and resumes the *Carry* at the command **FRONT**. The guard on the right flank of the Color guard does not execute *Eyes Right.*

d. During ceremonies, the Color guard remains at *Right Shoulder Arms* except when executing *Present Arms* (as specified in subparagraph f).

e. When not participating in a ceremony and a situation occurs that warrants a *Salute* by the organizational Color, the Color sergeant commands ***Color,*** **SALUTE**. The return to the *Carry* is made at the command ***Carry,*** **COLOR**.

f. When in formation with the Color company, and not during a ceremony, the Color bearers execute *At Ease* and *Rest,* keeping the staffs of the Colors vertical. The Color guard executes *Right Shoulder Arms, Order Arms,* and *Present Arms* with the Color company. During ceremonies when the Colors are not forward and remarks are to be made, the Color guards and Color bearers execute *Order Arms* and *Parade Rest* on command of the Color company commander. During any ceremony when the units are *At Ease,* the Color guard and Color bearers are at *Parade Rest.*

g. The uniform for Color guards should be the same as prescribed for participating troops.

15-5. RECEIVING OR DISMISSING THE COLORS BY THE COLOR GUARD

The Color guard uses the following procedures when receiving or dismissing the Colors.

a. When receiving uncased Colors on display in the commander's office, the Color guard is positioned in a single rank facing the Colors. The Color sergeant commands ***Present,*** **ARMS** and ***Order,*** **ARMS**. On completion of *Order Arms,* the Color bearers (without command) secure the Colors. The Color guard files outside (guard, National Color, organizational Color, guard) and re-forms in a line formation. The Color guards execute *Right Shoulder Arms* and the Color bearers assume the *Carry Position.*

b. To dismiss the Colors, the procedures are basically the same except that the Colors are placed back in their stands before executing *Present Arms.*

15-6. CASING AND UNCASING THE COLORS

Use the following procedures to case and uncase the Colors.

a. The Color guard forms in a line formation with the cased Colors at the *Carry Position (Order Position* when indoors). The command sergeant major (or his direct representative) positions himself six steps in front of and centered on the formation. He then commands ***Sling,*** **ARMS**.

(1) The Color guards immediately adjust their slings and assume *Sling Arms.* The command sergeant major commands **POST**. The Color guards face to the *Half Left (Right)* in marching, take four steps, halt, and execute *About Face.*

(2) The command sergeant major then directs **UNCASE THE COLORS**. The Color bearers lower the Colors (same as *Present Guidon).* The two guards move forward and untie and uncase the Colors. The Color bearers unfurl and immediately return the Colors to the *Carry (Order) Position.* While the Colors are being unfurled, the guards fold the cases and secure them in their left hand.

(3) When the Colors are in the *Carry Position,* the command sergeant major commands ***Present,*** **ARMS**. The command sergeant major, Color guards, and the organizational Color salute.

(4) The command sergeant major commands ***Order,*** **ARMS**, and then commands **POST**. On the command of execution **POST**, the Color guards place the folded canvas cases inside the cartridge belts (center rear) of the Color bearers. The Color guards assume their original positions, adjust their slings, and return to *Right Shoulder Arms.*

(5) If the command sergeant major or his representative is not present, the senior Color sergeant gives the necessary commands.

b. To case the Colors, the procedures are basically the same except *Present Arms* is given before the Colors are lowered.

c. When casing or uncasing the Colors with the command present, the commander directs **UNCASE (CASE) THE COLORS**. The command sergeant major and Color guards execute the movement (as previously stated) except that they execute *Present Arms* and *Order Arms* with the Color company. When the Colors are uncased and returned to the *Carry Position,* the commander directs **BRING YOUR UNITS TO PRESENT ARMS**. After the units have executed this directive, he then directs **BRING YOUR UNITS TO ORDER ARMS.**

d. If the Colors are to be cased or uncased during the receiving or dismissing by the Color company, the Color guards execute *Present Arms* and *Order Arms* on command of the company commander.

e. The command sergeant major uncases the organizational Color when it is displayed by itself. He may also assist the Color guards when uncasing more than two Colors.

15-7. RECEIVING OR DISMISSING THE COLORS BY COLOR COMPANY

The designated Color company receives and dismisses the Colors as follows:

a. The Color company should receive the Colors before it forms with the battalion.

b. The Color company forms at *Attention* with the commander facing to the front as the Colors near the Color company.

c. The Color guard, guided by the senior sergeant, approaches from the front and halts 10 steps from the company commander.

d. The company commander then faces about and commands ***Present,*** **ARMS**, faces the Colors and salutes. He terminates his *Salute,* faces about, and commands ***Order,*** **ARMS.**

e. The specialists or privates of the Color guard execute *Present Arms* and return to *Right Shoulder Arms* on the commands of the company commander *(Present Arms* and *Order Arms).*

f. The senior sergeant then marches the Color guard to its position within the company formation. If the company is in column formation, the Color guard forms at the rear of the company. When the company is in line or mass formation, the Color guard forms at the left of the company. The Color company may join the battalion before the battalion forms at the ceremony site or join with the battalion at the ceremony site. When the Color company joins the battalion, the senior Color sergeant marches the Color guard to its appropriate post in the battalion formation.

g. The Color guard is dismissed at the conclusion of the ceremony. This can occur in the vicinity of the ceremony site, in the Color company area, or at the battalion headquarters. At the designated area, the senior sergeant marches and halts the Color guard 10 steps in front of and facing the commander of the Color company. The actions for dismissing the Colors are the same as receiving the Colors. After being dismissed, the Color guard marches to the office, headquarters, or tent of the commanding officer.

h. The Colors are received and dismissed from organizations smaller than a company, such as a funeral escort, in a similar manner.

i. Casing and uncasing the Colors may be scheduled in conjunction with receiving and dismissing the Colors.

15-8. POSTING AND RETIRING THE COLORS

Formal assemblies conducted indoors begin with the presentation of the Colors, referred to as posting the Colors, and end with the retirement of the Colors. The following instructions outline the procedures for posting and retiring the Colors, with a head table and without head table. Since indoor areas vary in size, configuration, and intended purpose, these instructions do not apply to all situations. Therefore, persons planning an indoor ceremony can modify these instructions based on their specific floor plan.

a. **Posting the Colors.** The Color guard forms outside the entrance to the dining area, auditorium, or meeting hall. The audience is directed to stand until the Colors are posted. If the playing of the National Anthem (or other appropriate music) and the invocation are scheduled, the audience will remain standing until they are completed.

(1) When the arrangements include a head table, the Color guard enters in a line formation, preferably, or forms in a line immediately inside the room and moves to a position centered on and facing the head table.

(a) When the Colors arrive at the predesignated position, the Color sergeant commands ***Colors,*** **HALT**, and ***Present,*** **ARMS**; and reports *"**The colors are present.**"* The host acknowledges the report and directs **POST THE COLORS**. The area should be arranged to allow adequate space for the Color guard to move between the head table and the flag stand.

(b) The Color sergeant then commands ***Order,*** **ARMS;** ***Right,*** **FACE**; and ***Forward,*** **MARCH.** On the command of execution **MARCH**, the Color guard marches to the rear of the head table (Figure 15-1, page 15-8).

(c) Once the Color guard is centered on the flag stand, they mark time and the Color sergeant commands ***Colors,*** **HALT** and ***Right,*** **FACE**. The Color guard should approach the flag stands from the right to position the National Color bearer in front of the flag holder on the right, facing the audience.

(d) The Color bearers, without command, place the colors in the stand.

(e) When the Colors are in the stand, the color sergeant commands ***Present,*** **ARMS** and ***Order,*** **ARMS**. The guards return to *Right Shoulder Arms;* the Color sergeant commands ***Left,*** **FACE** and ***Forward,*** **MARCH**; and the Color guard exits the area.

(2) When a head table is not used, the Color guard enters and moves to a predesignated position centered on and facing the audience. This may require the Color guard to move in a column and use *Facing* movements. The movement must be planned so that the National Color is always on the right when in line and is leading when in column.

(a) When the Colors arrive at the predesignated position, the Color sergeant commands ***Colors,*** **HALT;** ***Left (Right),*** **FACE;** and ***Present,*** **ARMS**. Any scheduled music or the Pledge of Allegiance occurs at this time. The Color sergeant then commands ***Order,*** **ARMS.**

(b) The Color sergeant commands ***Right (Left),*** **FACE** and ***Forward,*** **MARCH**. On the command of execution **MARCH**, the Color guard marches to the flag stand where the actions of the Color guard are the same as previously described.

b. **Retiring the Colors.** The audience is directed to stand for the retiring of the Colors.

(1) When a head table is used, the Color sergeant moves the Color guard to the head table.

(a) The Color sergeant commands ***Color guard,*** **HALT;** ***Present,*** **ARMS**, and reports to the host, ***"Sir, request permission to retire the colors."*** The host acknowledges the report and directs **RETIRE THE COLORS**.

(b) The Color sergeant commands ***Order,*** **ARMS;** ***Right,*** **FACE;** ***Forward,*** **MARCH**; and moves the Color guard until they are centered on the flag stand where they mark time.

(c) The Color sergeant commands ***Color guard,*** **HALT;** ***Right,*** **FACE;** ***Present,*** **ARMS**; and ***Order,*** **ARMS**. Upon completion of *Order Arms* the color bearers, without command, retrieve the colors and assume the *Carry Position.*

(d) The Color sergeant commands ***Left,*** **FACE** and ***Forward,*** **MARCH**. The Color guard exits the area.

(2) When the head table is not used, the Color guard moves directly to the flag stands where the Colors are retrieved. The Color guard exits as previously described.

NOTE: These procedures will vary when the command sergeant major is in charge of the Colors during a formal dining-in.

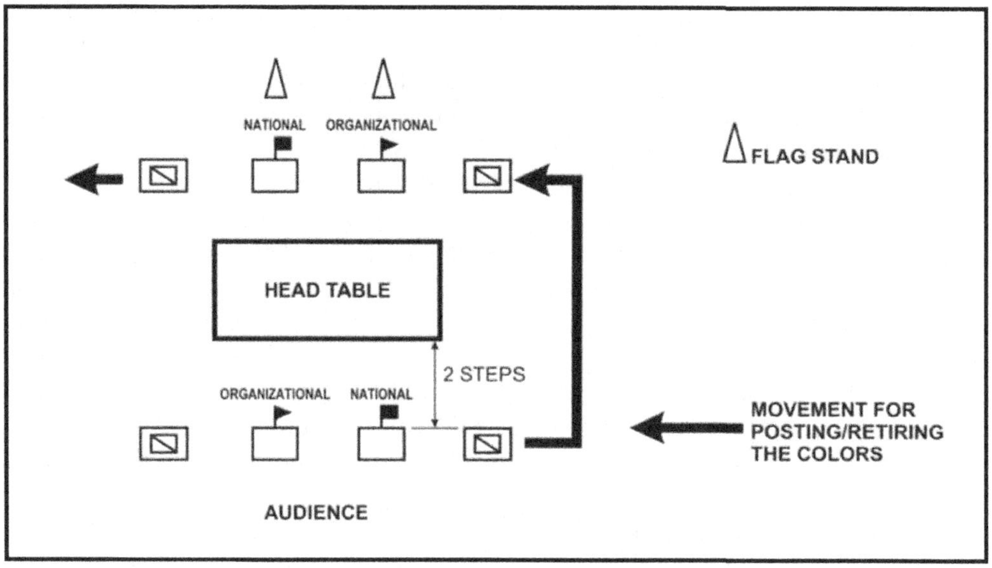

Figure 15-1. Posting and retiring the Colors.

15-9. MOVEMENT OF THREE-MAN COLOR GUARD TO THE REAR

To face a three-man Color guard to the rear, the command is ***Colors Reverse,*** **MARCH**, and each man simultaneously executes the following movements:

a. Number 1 faces left while marking time, takes two steps forward, and faces to the left while marking time.

b. Number 2 takes one full step forward, executes *About Face* while marking time, and takes one full step in the new direction, marking time.

c. Number 3 takes one full step and two half steps forward, faces to the right while marking time, takes two full steps, faces to the right in marching, takes two full steps, and marks time.

d. When all are abreast of each other, they step off together or halt, as commanded (Figure 15-2).

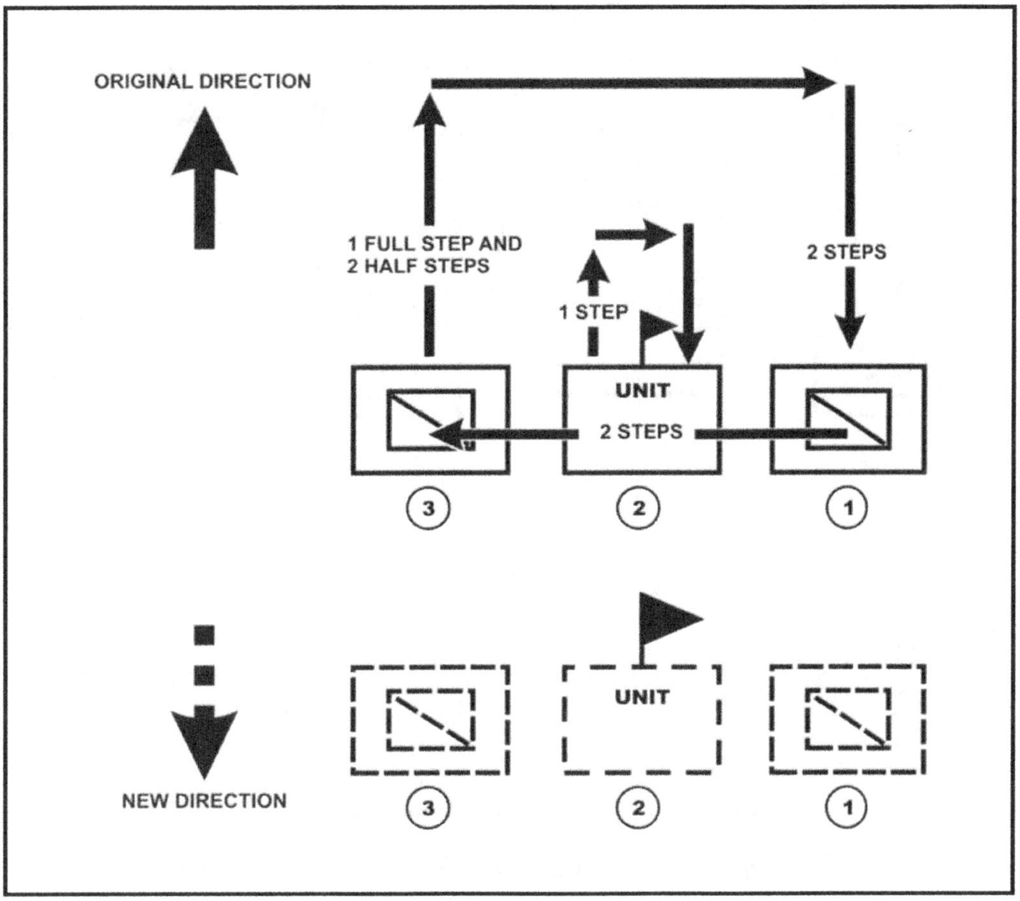

Figure 15-2. Movement of Three-man Color Guard to the Rear.

15-10. MOVEMENT OF FOUR-MAN COLOR GUARD TO THE REAR

To face a four-man Color guard to the rear, the command is ***Colors reverse,*** **MARCH**. At the command **MARCH**, each man simultaneously executes the following movements:

a. Number 1 takes two steps forward, faces to the left in marching, takes one full step and three half steps, faces to the left in marching, and takes two steps in the new direction.

b. Number 2 faces to the left in marching and faces to the left while marking time.

c. Number 3 takes one step forward, faces to the right in marching, takes two half steps, faces to the right in marching, and takes one step in the new direction.

d. Number 4 takes one step forward, faces to the right in marching, takes one full step and three half steps, faces to the right in marching, and takes one step forward in the new direction.

e. Numbers 2, 3, and 4 mark time after completing their movements until all men are abreast, then step off together or halt, as the situation dictates (Figure 15-3, page 15-10).

NOTE: This command may be given while marching, in which case the commands are given when the left foot strikes the marching surface.

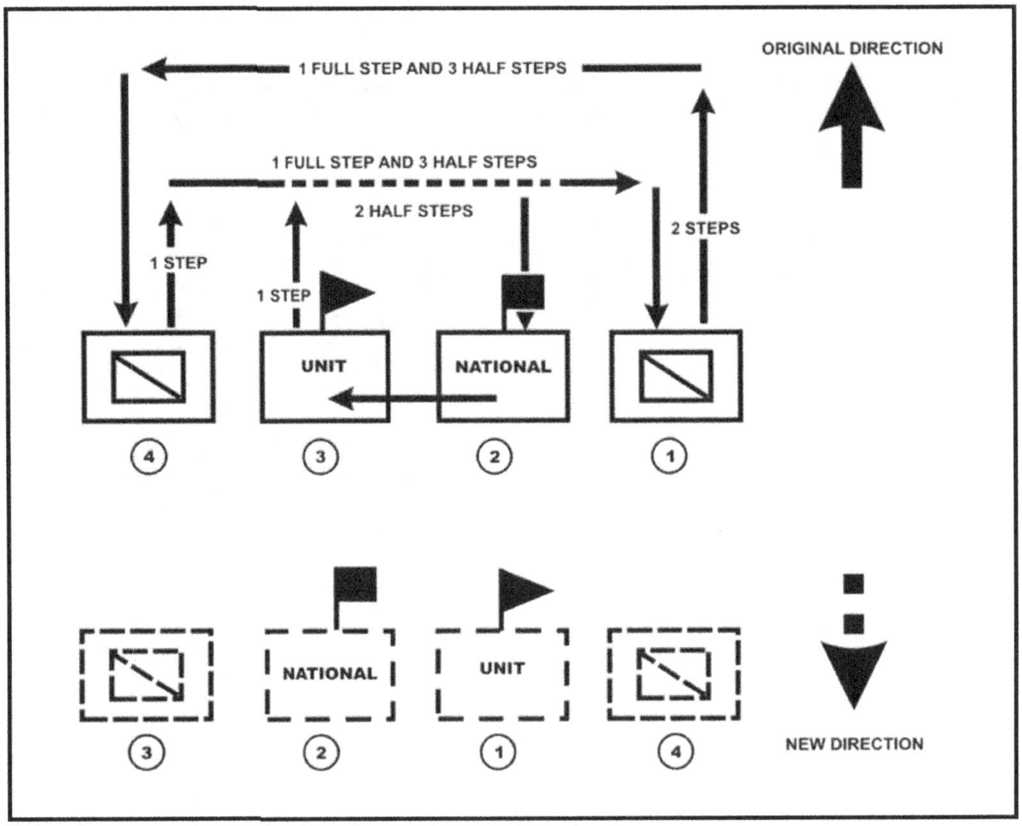

Figure 15-3. Movement of four-man Color guard to the rear.

NOTE: *Left About* may be used in lieu of *Colors Reverse*. The command for this movement is ***Left About*, MARCH**. The Color sergeant is the pivot man for the movement. On the command of execution, the Color sergeant marches in place and turns to the left. Other members shorten their step and turn in an arc keeping abreast of each other and maintaining alignment. When the movement has been completed, each man marches in place until the command **HALT** or ***Forward*, MARCH** is given. This movement may be executed from the **HALT** or while marching.

15-11. MOVEMENT OF FIVE-MAN COLOR GUARD TO THE REAR

To face a five-man Color guard to the rear, the command is ***Colors Reverse*, MARCH**. At the command **MARCH**, each man simultaneously executes the following movements:

a. Number 1 takes four steps forward, faces to the left in marching, takes four full steps forward, faces to the left in marching, takes four full steps in the new direction, and marks time.

b. Number 2 faces left while marking time, takes two steps forward, and faces to the left while marking time.

c. Number 3 takes one full step forward, executes about face while marking time, takes one full step in the new direction, and marks time.

d. Number 4 takes one full step and two half steps, faces to the right while marching, takes two full steps, faces to the right in marching, takes two full steps, and marks time.

e. Number 5 takes three full steps forward, faces to the right in marching, takes four full steps forward, faces to the right in marching, takes three full steps forward, and marks time.

f. Numbers 2, 3, 4, and 5 mark time after completing their movements.

g. When all men are abreast of each other, they step off together or halt, as the situation dictates (Figure 15-4).

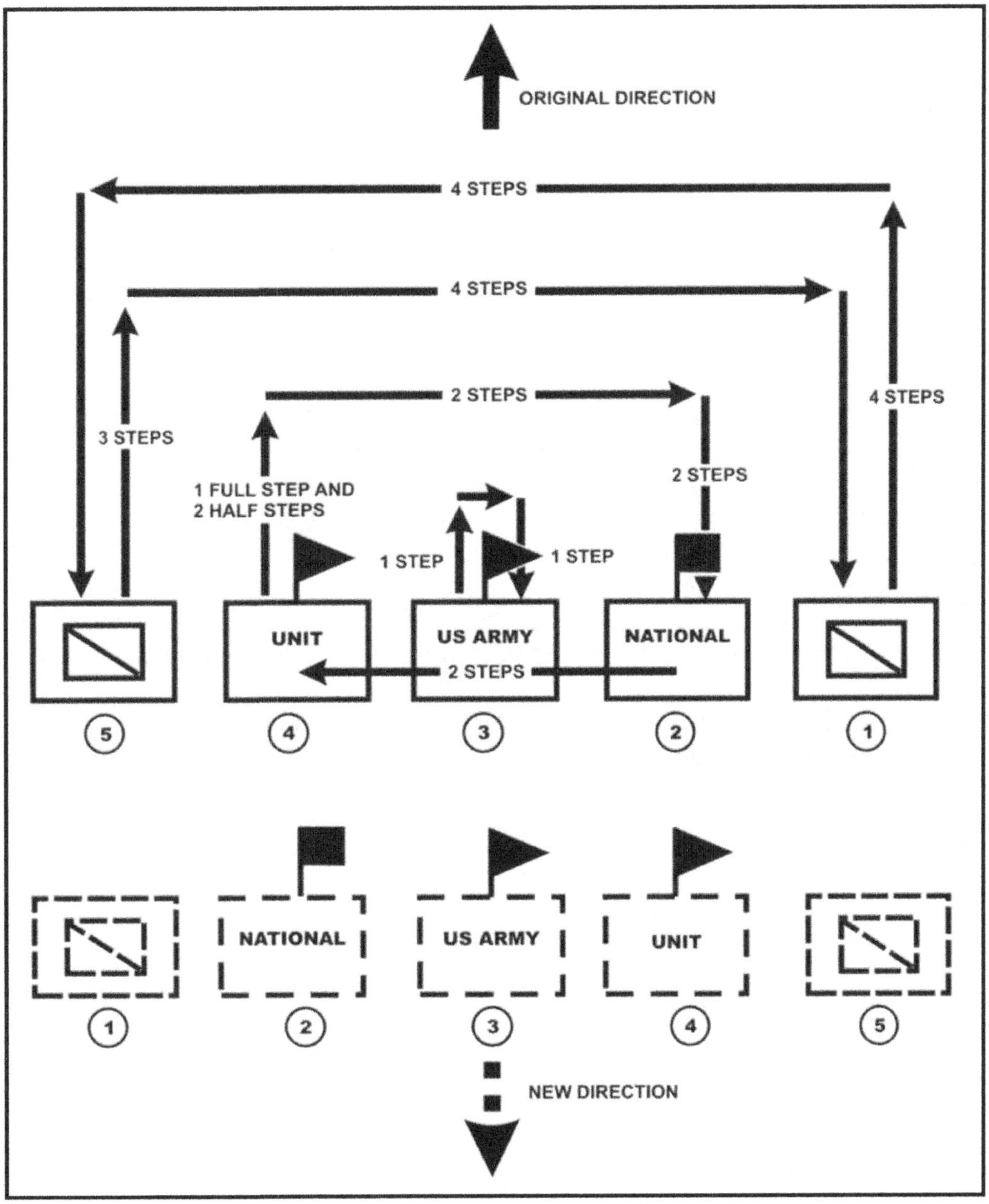

Figure 15-4. Movement of five-man Color guard to the rear.

NOTE: *Left About* may be used in lieu of *Colors Reverse*. The command for this movement is ***Left about,*** **MARCH**. The Color sergeant is the pivot man for the movement. On the command of execution, the Color sergeant marches in place and turns to the left. Other members shorten their step and turn in an arc keeping abreast of each other and maintaining alignment. When the movement has been completed, each man marches in place until the command **HALT** or ***Forward,*** **MARCH** is given. This movement may be executed from the **HALT** or while marching.

15-12. MOVEMENT OF SIX-MAN COLOR GUARD TO THE REAR

To face a six-man Color guard to the rear, the command is ***Colors Reverse,*** **MARCH**. At the command **MARCH**, each man simultaneously executes the following movements:

a. Number 1 takes five steps forward, faces to the left in marching, takes six full steps forward, faces to the left in marching, takes five full steps in the new direction, and marks time.

b. Number 2 faces left while marking time, takes three full steps forward, and faces to the left in the new direction, while marking time.

c. Number 3 takes two full steps forward, faces to the left while marching, takes two steps forward, faces to the left while marching, takes two full steps in the new direction, and marks time.

d. Number 4 takes one full step forward, faces to the right while marching, takes two half steps forward, faces to the right in marching, takes one full step in the new direction, and marks time.

e. Number 5 takes three full steps forward, faces to the right in marching, takes three full steps forward, faces to the right in marching, takes three full steps in the new direction, and marks time.

f. Number 6 takes four full steps forward, faces to the right in marching, takes six full steps forward, faces to the right in marching, takes four full steps in the new direction, and marks time.

g. Numbers 2, 3, 4, 5, and 6 mark time after completing their movement.

h. When all men are abreast of each other, they step off together or halt, as the situation dictates (Figure 15-5).

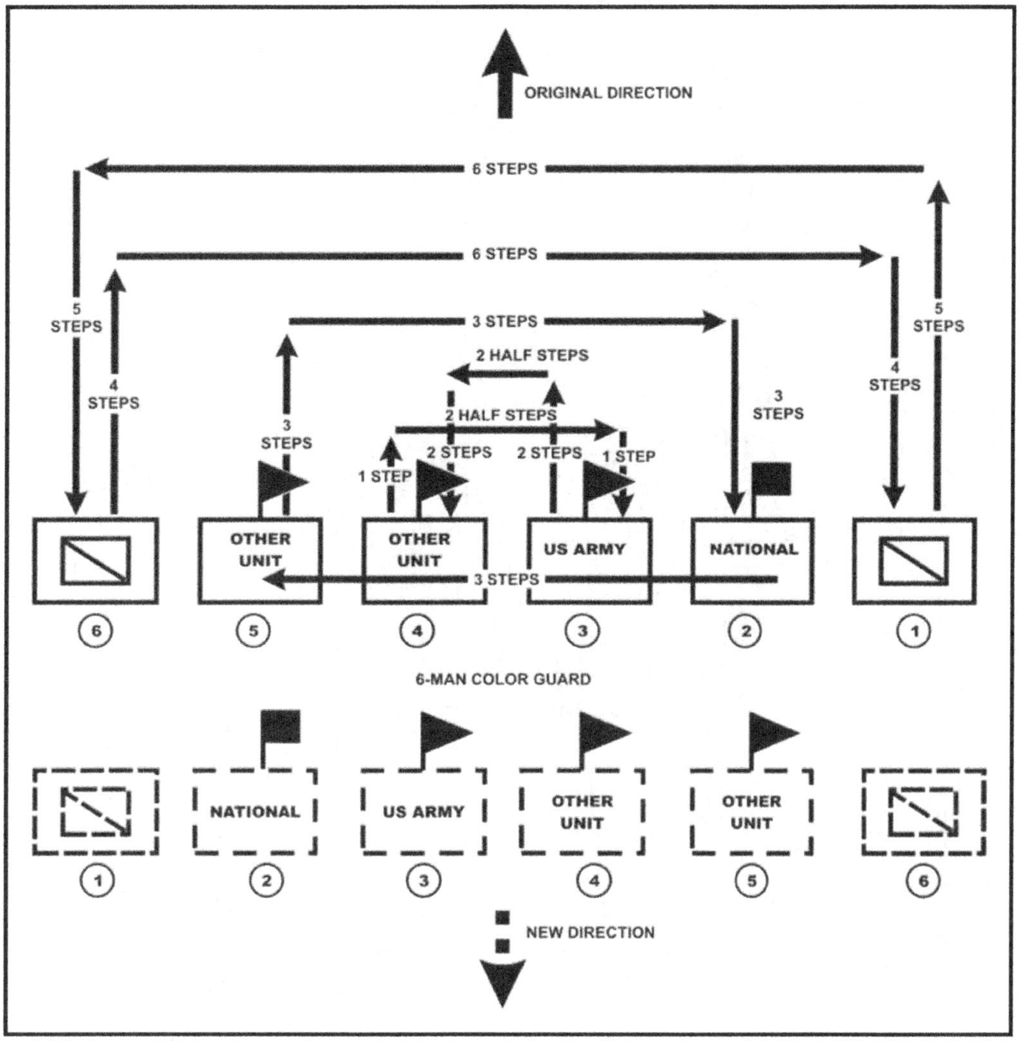

Figure 15-5. Movement of six-man Color guard to the rear.

15-13. POSITION OF THE COLORS AT THE ORDER

At the *Order*, rest the ferrule of the staff on the marching surface touching the outside of the right footgear opposite the ball of the right foot. Hold the staff in the right hand with the back of the hand facing outward and the fingers wrapped around the staff. Rest the staff against the hollow of the shoulder (Figure 15-6, page 15-14).

Figure 15-6. Position of the Colors at the Order.

15-14. POSITION OF THE COLORS AT THE CARRY

At the *Carry,* rest the ferrule of the staff in the socket of the sling. The socket is below the waist and adjusted to ensure that the finials of all Colors are of equal height (Figure 15-7). Grasp the staff with the right hand (even with the mouth) and incline it slightly to the front with the left hand securing the ferrule in the socket. The left hand may be positioned immediately below the right hand to more firmly secure the Colors on windy days.

Figure 15-7. Position of the Colors at the Carry.

15-15. POSITION OF THE COLORS AT PARADE REST

Parade Rest with the Colors is executed with staffs vertical. The *Order* and *Parade Rest* are executed with the Color company except during ceremonies when the Colors remain at the *Carry Position* (Figure 15-8).

Figure 15-8. Position of the Colors at Parade Rest.

15-16. POSITION OF THE ORGANIZATIONAL COLOR AT COLOR SALUTE

This position is assumed from the carry by slipping the right hand upward about 4 inches and then thrusting the arm forward shoulder high and horizontal to the marching surface forming an approximate 45-degree angle. When the Colors salute with troops who execute *Present Arms* from the *Order*, the Color bearers assume the position of *Carry* at the command **ARMS** and then execute the Color *Salute* (Figure 15-9). When casing or uncasing Colors indoors (at the *Order*), the organizational Color bearer salutes (dips) by slipping his right hand upward and grasping the staff firmly, with the forearm horizontal and the staff thrust forward until the arm is fully extended.

Figure 15-9. Position of the organizational Color at Color Salute.

This page intentionally left blank.

CHAPTER 16
SALUTE BATTERY

Ceremonies did not originally use field artillery as a salute battery. In the 19th century, branches of service were lined up from right to left: infantry, field artillery, and cavalry, respectively. Field artillery lined up their guns but did not fire them. When honors began to be fired in ceremonies, the guns were moved out of the formation to avoid injury to other elements of the formation.

16-1. PREPARATION

The salute battery should arrive at the ceremonial site early enough to position their pieces before the arrival of units on the ready line.

a. The pieces are positioned downwind from the ceremony site with the muzzles pointing away from the reviewing stand. When possible, the opened trails are two steps to the rear of the final line and the number six piece is positioned twelve steps from the left flank unit. Other pieces are positioned about four steps apart (measure from inside wheels) and on line with number six.

b. Once the pieces are in position, the battery is formed to the left of the number one piece and in front of the line so that the number six and three sections are on the final line (Figure 16-1).

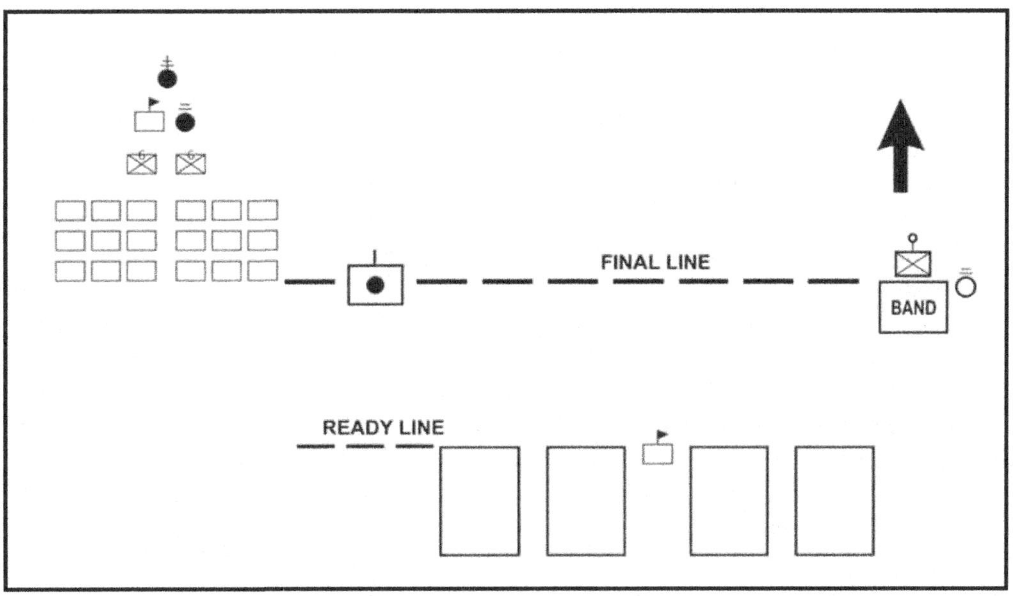

Figure 16-1. Position of salute battery and personnel before the review.

16-2. SEQUENCE OF EVENTS AND INDIVIDUAL ACTIONS

The sequence of events and individual actions necessary to conduct a salute battery are described herein.

a. **Formation of Troops.** On hearing the band sound *Attention,* the battery commander faces about, commands ***Battery,*** **ATTENTION**, and again faces about.

(1) On completion of the adjutant's call, the battery commander *faces about* and commands ***Right,*** **FACE**. He then commands ***Section, POST***.

(2) The section chiefs in reverse sequence (6, 5, 4, 3, 2, 1) command ***Forward,*** **MARCH** or ***Column Half Right,*** **MARCH**. The sections march forward along the final line and halt when centered on the piece.

(3) The battery commander faces to the left and commands (himself and his staff) ***Column Half Right,*** **MARCH**. They march forward (inclining as necessary) so that when halted the battery commander is six steps from and centered on the firing battery. The section chief and the counter should be centered and three steps to the rear of the commander. The executive officer is on line with the guidon bearer.

(4) When the battery commander has halted his staff, he faces to the right and commands ***Left,*** **FACE.** Then he commands ***Dress right,*** **DRESS**; ***Ready,*** **FRONT**; and ***Parade,*** **REST** (for large reviews, the command **AT EASE** may be substituted for ***Parade,*** **REST**). The commander then *faces about* and executes *Parade Rest.*

b. **Presentation of Honors.** When the commander of troops has directed *Order Arms* (adjutant has taken his post), the battery commander faces about and commands ***Order,*** **ARMS.** He then directs **FIRE THE SALUTE** and faces about.

(1) The battery commander quietly commands (himself and the guidon bearer) ***Present,*** **ARMS** and ***Order,*** **ARMS**, as appropriate.

(2) The executive officer, counter, and chief of the firing battery face about. The executive officer commands ***Cannoneers,*** **POST**. The cannoneers immediately move to their firing positions.

(3) The executive officer commands ***With blank ammunition*** (pause), **LOAD**. He then commands **STAND BY, NUMBER ONE**. As each piece is ready to fire, the section chief raises his arm (nearest the executive officer) and looks at the executive officer.

(4) When all section chiefs have indicated they are ready (arm raised), the executive officer raises his right arm and looks at the commander of troops. The commander of troops takes the raised arm as the signal that the battery is ready to fire. He then faces about and commands (himself and his staff) ***Present,*** **ARMS**.

(5) As the commander's hand touches his headgear, the executive officer signals the first piece to fire. The executive officer drops his arm to a horizontal position and then raises it back to a vertical position for each round. Rounds are fired at three-second intervals for normal honors and five-second intervals for funerals (AR 600-25).

(6) During the *Salute,* as the second to last round is fired, the counter faces about and announces **ELEVEN**. As the next to last round is fired, the chief of the firing battery *Faces About* and announces **TWELVE**. This is the signal for the executive officer to fire one more round.

(7) During the firing, each section immediately reloads after each round is fired and is prepared to fire in case of a misfire. When the last round has been fired, the executive officer signals **CEASE FIRE** and then signals **UNLOAD**.

(8) When all pieces are unloaded and the cannoneers are at *Attention,* the executive officer commands to the rear of the piece, **FALL IN** (cannoneers automatically fall in at *Dress Right);* he then commands ***Ready,*** **FRONT** and ***Parade,*** **REST**. When all sections are back in position on the final line, the executive officer faces about.

c. **Inspection.** As the reviewing party approaches piece number six, the battery commander faces about and commands ***Battery,*** **ATTENTION**. He then faces to the front and (over his right shoulder) commands ***Eyes,*** **RIGHT**.

(1) The battery commander and executive officer execute *Eyes Right* and *Salute.* The guidon bearer executes *Eyes Right* and *Present Guidon.* All others execute *Eyes Right* and follow the reviewing officer, turning the head until facing to the front.

(2) When the reviewing party has passed the last piece, the battery commander commands ***Order,*** **ARMS** for himself, the executive officer, and the guidon bearer. He then faces about. When the reviewing party has cleared the right rear of his unit, he then commands ***Parade,*** **REST**, faces about, and assumes *Parade Rest.*

d. **Honors to the Nation.** The battery executes appropriate movements on command or directives by the commander of troops.

e. **Remarks.** The battery executes appropriate movements on command or directives by the commander of troops.

f. **March in Review.** On the command or directive to **PASS IN REVIEW**, the battery commander faces about and commands ***Parade,*** **REST.** He then faces about and assumes *Parade Rest.*

g. **Conclusion.** When the band begins the eight-bar drum beat (before "The Army Goes Rolling Along"), the battery commander faces about and commands ***Battery*** **ATTENTION.** He then faces about. When the band has finished playing the music, the battery commander faces about and commands **REST**.

This page intentionally left blank.

CHAPTER 17

CIVILIAN PARTICIPATION IN MILITARY CEREMONIES

This chapter serves as a guide for commanders in planning and conducting military ceremonies with civilian participation.

17-1. HISTORY

Since its inception at the beginning of the American Revolution, the U.S. Army has always seen the importance of civilian control of the military and participation of civilians in military activities. As a result, military ceremonies are frequently held to honor civilians.

17-2. TYPES OF CEREMONIES

Civilians may participate in three general categories of military ceremonies:

- Presenting awards.
- Receiving awards.
- Being honored.

a. **Civilians Presenting Awards.** This category includes occasions when an award is made to another civilian and occasions when awards are presented to military personnel by a civilian. These ceremonies are conducted as previously outlined with the following exceptions:

(1) The host or reviewing officer is accompanied to the reviewing stand by the civilian. The reviewing officer may relinquish the place of honor to the civilian.

(2) The reviewing officer accompanies the civilian throughout the sequence of events.

(3) Because of lack of familiarity with military ceremonies, it may be necessary to brief the civilian before and during the conduct of the ceremony.

b. **Civilians Receiving Awards.** Conduct of these ceremonies should be guided by paragraph 10-4 with the following exceptions:

(1) Civilians should be initially positioned to the left of the reviewing stand and not marched forward with the military personnel, if any, to receive awards.

(2) When the Colors are brought forward, civilians move to a position five steps in front of and centered on the Colors. When military and civilians receive an award at the same time, they are aligned from right to left with the highest award on the right.

(3) When the awardees are posted, they remain in the same order. If there are no military awardees, the civilians move to the left of the reviewing stand.

(4) When this ceremony is conducted for presenting posthumous awards, other types of awards are not incorporated. A carefully selected escort officer should accompany relatives of persons receiving posthumous awards.

NOTE: If there is only one awardee, he may be the reviewing official.

c. **Reviews in Honor of Civilians.** Government employees, private citizens, and local, state, national, or foreign officials may be honored by a review. (Consult AR 600-25 for special honors due certain officials.) Other civilians may be honored, as noted above, by participating in the place of or with the reviewing officer.

APPENDIX A

SALUTING

The origin of the Hand Salute is uncertain. Some historians believe it began in late Roman times when assassinations were common. A citizen who wanted to see a public official had to approach with his right hand raised to show that he did not hold a weapon. Knights in armor raised visors with the right hand when meeting a comrade. This practice gradually became a way of showing respect and, in early American history, sometimes involved removing the hat. By 1820, the motion was modified to touching the hat, and since then it has become the Hand Salute used today.

A-1. WHEN TO SALUTE

Army personnel in uniform are required to salute when they meet and recognize persons entitled (by grade) to a salute except when it is inappropriate or impractical (in public conveyances such as planes and buses, in public places such as inside theaters, or when driving a vehicle).

a. A salute is also rendered—

- When the United States National Anthem, "To the Color," "Hail to the Chief," or foreign national anthems are played.
- To uncased National Color outdoors.
- On ceremonial occasions as prescribed in Part Two, Ceremonies.
- At reveille and retreat ceremonies, during the raising or lowering of the flag.
- During the sounding of honors.
- When the Pledge of Allegiance to the U.S. flag is being recited outdoors.
- When turning over control of formations.
- When rendering reports.
- To officers of friendly foreign countries.

b. Salutes are not required when—

- Indoors, except when reporting to an officer or when on duty as a guard.
- Addressing a prisoner.
- Saluting is obviously inappropriate. In these cases, only greetings are exchanged. (Example 1: A person carrying articles with both hands, or being otherwise so occupied as to make saluting impracticable, is not required to salute a senior person or return the salute to a subordinate.)
- Either the senior or the subordinate is wearing civilian clothes.

c. In any case not covered by specific instructions, the salute is rendered.

A-2. REPORTING INDOORS

When reporting to an officer in his office, the soldier removes his headgear, knocks, and enters when told to do so. He approaches within two steps of the officer's desk, halts, salutes, and reports, ***"Sir (Ma'am), Private Jones reports."*** The salute is held until the report is completed and the salute has been returned by the officer. When the business is completed, the soldier salutes, holds the salute until it has been returned, executes the appropriate facing movement, and departs. When reporting indoors under arms, the procedure is the same except that the headgear is not removed and the soldier renders the salute prescribed for the weapon with which he is armed.

NOTES: 1. The expression "under arms" means carrying a weapon in your hands by a sling or holster.
2. When reporting to a noncommissioned officer, the procedures are the same, except no salutes are exchanged.

A-3. REPORTING OUTDOORS

When reporting outdoors, the soldier moves rapidly toward the officer, halts approximately three steps from the officer, salutes, and reports (as when indoors). When the soldier is dismissed by the officer, salutes are again exchanged. If under arms, the soldier carries the weapon in the manner prescribed for saluting.

A-4. SALUTING PERSONS IN VEHICLES

The practice of saluting officers in official vehicles (recognized individually by grade or identifying vehicle plates and or flags) is considered an appropriate courtesy. Salutes are not required to be rendered by or to personnel who are driving or riding in privately owned vehicles except by gate guards, who render salutes to recognized officers in all vehicles unless their duties make the salute impractical. When military personnel are drivers of a moving vehicle, they do not initiate a salute (AR 600-25).

A-5. OTHER SALUTES

Other instances when saluting is an issue are discussed herein.

a. **In Formation.** Individuals in formation do not salute or return salutes except at the command ***Present,*** **ARMS.** The individual in charge salutes and acknowledges salutes for the entire formation. Commanders of organizations or detachments that are not a part of a larger formation salute officers of higher grade by bringing the organization or detachment to attention before saluting. When in the field under battle or simulated battle conditions, the organization or detachment is not brought to attention. An individual in formation at ease or at rest comes to attention when addressed by an officer. (Refer to paragraph 3-5f, for more information on the hand salute).

b. **Not in Formation.** On the approach of an officer, a group of individuals not in formation is called to *Attention* by the first person noticing the officer, and all come sharply to *Attention* and salute. This action is to be taken at approximately 6 paces away from the officer, or the closest point of approach. Individuals participating in games, and members of work details, do not salute. The individual in charge of a work detail, if not actively engaged, salutes and acknowledges *Salutes* for the entire detail. A unit resting alongside a road does not come to *Attention* upon the approach of an officer; however, if the officer addresses an individual (or group), the individual (or group) comes to *Attention* and remains at *Attention* (unless otherwise ordered) until the termination of the conversation, at which time the individual (or group) salutes the officer.

c. **Outdoors.** Whenever and wherever the United States National Anthem, "To the Color," "Reveille," or "Hail to the Chief" is played, at the first note, all dismounted personnel in uniform and not in formation face the flag (or the music, if the flag is not in view), stand at *Attention,* and render the prescribed *Salute.* The position of *Salute* is held until the last note of the music is sounded. Military personnel not in uniform will stand at *Attention* (remove headdress, if any, with the right hand), and place the right hand over the heart. Vehicles in motion are brought to a *Halt.* Persons riding in a passenger car or on a motorcycle dismount and salute. Occupants of other types of military vehicles and buses remain in the vehicle and sit at attention; the individual in charge of each vehicle dismounts and renders the *Hand Salute.* Tank and armored car commanders salute from the vehicle.

d. **Indoors.** When the National Anthem is played indoors, officers and enlisted personnel stand at *Attention* and face the music, or the flag if one is present.

NOTE: Narrators or printed programs can help inform spectators of appropriate responses.

A-6. SALUTING COLORS

National and organizational flags, which are mounted on flagstaffs equipped with finials, are called Colors. Military personnel passing an uncased National Color salute at six steps distance and hold the *Salute* until they have passed six steps beyond it. Similarly, when the uncased Color passes by, they salute when it is six steps away and hold the *Salute* until it has passed six steps beyond them.

NOTE: Small flags carried by individuals, such as those carried by civilian spectators at a parade, are not saluted. It is improper to salute with any object in the right hand or with a cigarette, cigar, or pipe in the mouth.

A-7. UNCOVERING

Officers and enlisted men under arms uncover only when—

- Seated as a member of (or in attendance on) a court or board.
- Entering places of divine worship.
- In attendance at an official reception.

Personnel remove their headdress indoors. When outdoors, military headdress is never removed, or raised as a form of salutation. When appropriate, civilians may be saluted in lieu of removing the headdress.

A-8. SALUTING UPON BOARDING NAVAL SHIPS

When Army personnel board U.S. Navy ships, either as an individual or as a unit leader, they salute according to naval procedures.

a. When boarding a naval ship, upon reaching the top of the gangway, face and salute the national ensign. After completing this salute, salute the officer of the deck who will be standing on the quarter deck at the head of the gangway. When saluting the officer of the deck, request permission to board, ***"Sir, Request permission to come aboard."*** The officer of the deck will return the salute.

b. When leaving the ship, render the same salutes in reverse order, and request permission to leave, ***"Sir, Request permission to go ashore."***

APPENDIX B
MANUAL OF ARMS—M4-SERIES CARBINE

The manual of arms movements for the M4-series carbine (Figure B-1) are nearly the same as those for the M16-series rifle. For ease of understanding, the movements are fully explained. The same rules that apply to the M16-series rifle apply to the M4-series carbine. Only conduct drills with the M4-series carbine with the standard handguards with the stock fully extended.

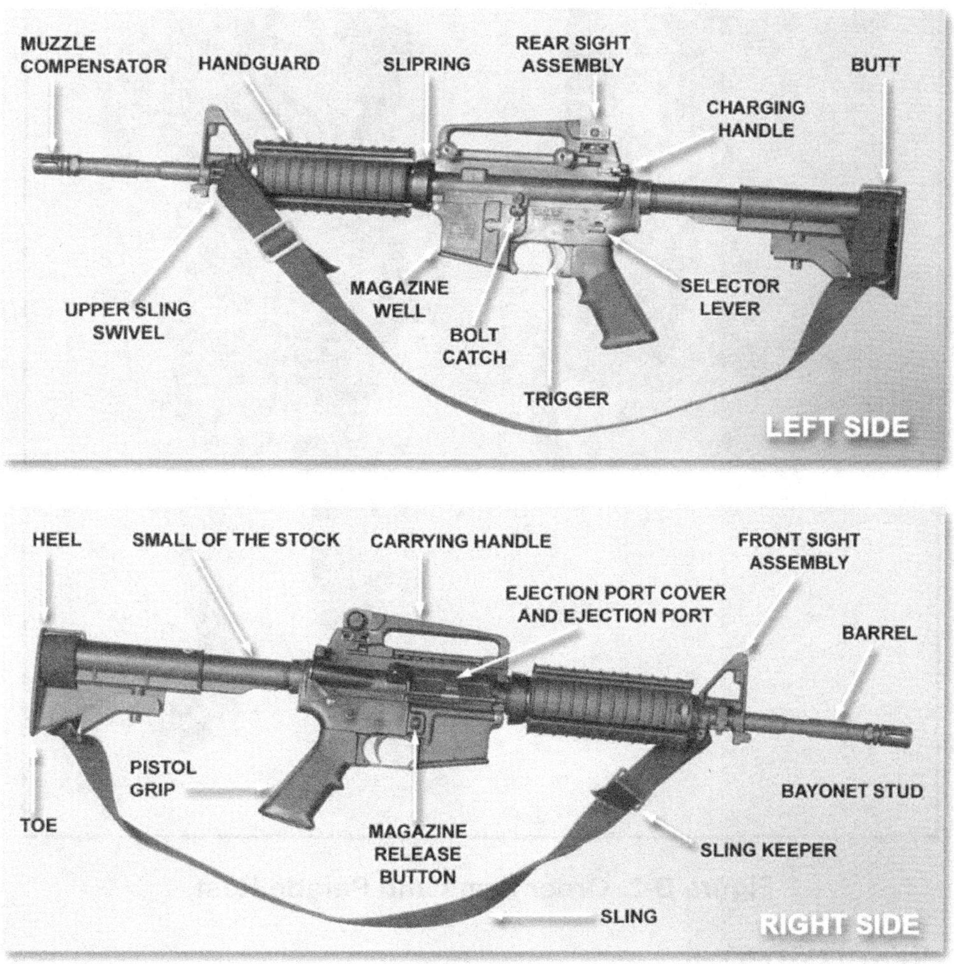

Figure B-1. Nomenclature, M4-series carbine.

B-1. ORDER ARMS

Assume *Order Arms* on the command **FALL IN** or from *Parade Rest* (Figure B-2, page B-2) on the command of execution **ATTENTION**. At *Order Arms*, maintain the *Position of Attention*. Grasp the weapon by cupping the butt of the carbine with the right hand, fingers joined with the thumb on the toe of the butt. Hold the carbine vertical with the magazine well to the front with the trailing edge of the right hand on line with the seam of the trousers.

Figure B-2. Order Arms and Parade Rest.

B-2. REST POSITIONS

Command and execute the rest positions the same as in individual drill with the following additions:

a. On the command of execution **REST** of ***Parade,*** **REST**, execute the movement as in individual drill without moving the right hand or the carbine.

b. On the command of execution **EASE** of ***Stand At,*** **EASE,** execute *Parade Rest* with the carbine except turn the head and eyes toward the person in charge of the formation.

c. On the command **AT EASE** or **REST**, keep the carbine in place as in *Parade Rest.*

B-3. PORT ARMS

To execute *Port Arms* use the following procedures:

a. *Port Arms* from *Order Arms* is a two-count movement (Figure B-3). The command is ***Port,*** **ARMS**. On the command of execution **ARMS**, reach across the body with the left hand and grasp the handguards just above the slipring. On count two, release the grasp of the right hand and raise the carbine diagonally across the body so that the carbine is about 4 inches from the waist. At the same time, grasp the small of the stock with the right hand. Hold the carbine diagonally across the body about 4 inches from the waist, right forearm horizontal, and the elbows close to the sides.

Figure B-3. Port Arms from Order Arms.

b. *Order Arms* from *Port Arms* is a two-count movement (Figure B-4, page B-4). The command is ***Order,*** **ARMS**. On the command of execution **ARMS**, release the grasp of the right hand and move the carbine to the side as in *Order Arms* with the carbine. At the same time, move the right hand to the side and then regrasp the butt of the carbine as in *Order Arms*. On count two, release the grasp of the left hand and return it sharply to the side and in the *Position of Attention*.

Figure B-4. Order Arms from Port Arms.

B-4. PRESENT ARMS

To execute *Present Arms* use the following procedures:

a. *Present Arms* from *Order Arms* is a three-count movement (Figure B-5). The command is ***Present,*** **ARMS.** On the command of execution **ARMS**, execute *Port Arms* in two counts. On count three, twist the carbine with the right hand so that the carbine is vertical and centered on the body about 4 inches from the waist with the magazine well to the front. Lower the carbine until the left forearm is horizontal and keep the elbows tight against the sides.

Figure B-5. Present Arms.

b. *Orders Arms* from *Present Arms* is a three-count movement. The command is ***Order,*** **ARMS.** On the command of execution **ARMS,** twist the rifle back to the *Port Arms* position. Counts two and three are executed the same as from *Port Arms.*

B-5. INSPECTION ARMS

To execute *Inspection Arms* use the following procedures:

a. *Inspection Arms* from *Order Arms* is a seven-count movement (Figure B-6, page B-6). The command is ***Inspection,*** **ARMS.**

(1) On the command of execution **ARMS**, execute *Port Arms* in the first two counts.

(2) On count three, move the left hand from the handguard and grasp the pistol grip, thumb over the lower portion of the bolt catch.

(3) On count four, release the grasp of the right hand, unlock the charging handle with the thumb, and sharply pull the charging handle to the rear with the thumb and forefinger. At the same time, apply pressure on the lower portion of the bolt catch, locking the bolt to the rear.

(4) On count five, without changing the grasp of the right hand, sharply push the charging handle forward until it is locked into position; then regrasp the rifle with the right hand at the small of the stock.

(5) On count six, remove the left hand, twist the carbine with the right hand so that the ejection port is skyward, regrasp the handguard with the left hand just forward of the slipring, and visually inspect the receiver through the ejection port.

(6) On count seven, with the right hand, twist the carbine so that the sights are up and come to *Inspection Arms.*

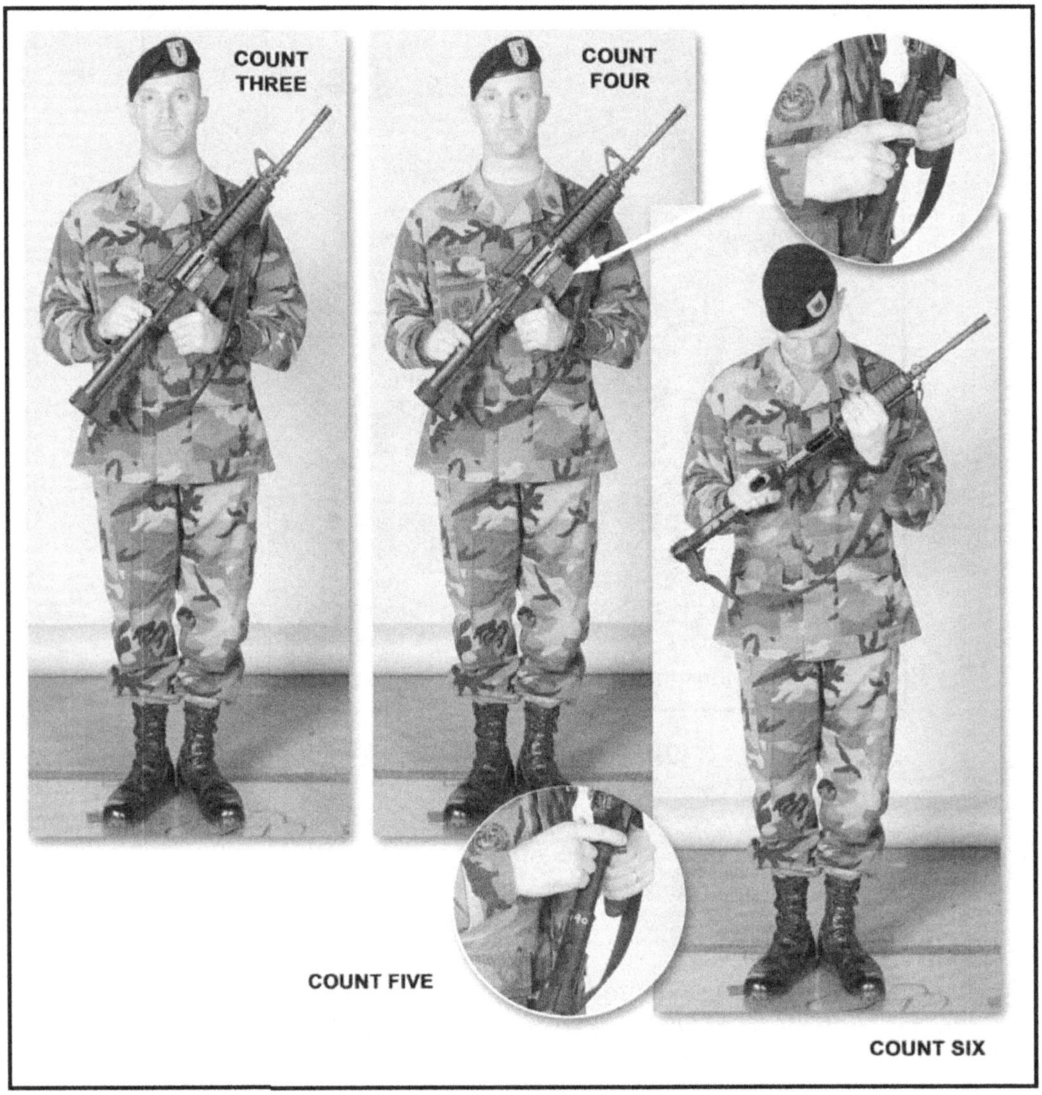

Figure B-6. Inspection Arms.

b. ***Ready, Port,*** **ARMS** is the only command given from *Inspection Arms.*

(1) On the command ***Ready,*** move the left hand and regrasp the carbine with the thumb and fingers forming a "U" at the magazine well and trigger guard, the thumb (without pressure) on the upper part of the bolt catch, fingertips placed below or under the ejection port cover.

(2) On the command ***Port,*** press the bolt catch and allow the bolt to go forward. With the fingertips, push upward and close the ejection port cover. Grasp the pistol grip with the left hand and place the left thumb on the trigger.

(3) On the command **ARMS,** pull the trigger and resume *Port Arms.* To return to *Order Arms* the command is ***Order,*** **ARMS.**

B-6. RIGHT SHOULDER ARMS

To execute *Right Shoulder Arms* use the following procedures:

a. *Right Shoulder Arms* from *Order Arms* is a four-count movement (Figure B-7). The command is ***Right Shoulder,*** **ARMS**.

(1) On the command of execution **ARMS**, execute count one the same as in *Port Arms.*

(2) On count two, execute in the same as *Port Arms* except move the right hand to grasp the butt of the carbine with the heel of the butt between the first two fingers with the thumb and forefinger touching.

(3) On count three (without moving the head), release the grasp of the left hand (without changing the grasp of the right hand), twist the carbine so that the sights are up, and place the weapon onto the right shoulder, moving the left hand to the small of the stock to guide the carbine to the shoulder. Keep the fingers and thumb (left hand) extended and joined with the palm turned toward the body. The first joint of the left forefinger touches the rear of the charging handle. Keep the left elbow down, and keep the right forearm horizontal with the right upper arm against the side and on line with the back.

(4) On count four, sharply move the left hand back to the left side as in the *Position of Attention.*

Figure B-7. Right Shoulder Arms.

b. *Order Arms* from *Right Shoulder Arms* is a three-count movement (Figure B-8). The command is ***Order,* ARMS**.

(1) On the command of execution **ARMS,** without moving the head and without changing the grasp of the right hand, press down quickly and firmly on the butt of the carbine with the right hand and twist the weapon (with the sights up), guiding it diagonally across the body and about 4 inches from the waist. Grasp the carbine with the left hand at the handguard just forward of the slipring.

(2) On count two, release the grasp of the right hand and move the rifle to the side as in *Order Arms*, then regrasp the butt with the right hand.

(3) On count three, release the grasp of the left hand and return it sharply to the side as in the *Position of Attention*, resuming *Order Arms*.

Figure B-8. Order Arms from Right Shoulder Arms.

B-7. LEFT SHOULDER ARMS

To execute *Left Shoulder Arms* use the following procedures:

a. *Left Shoulder Arms* from *Order Arms* is a four-count movement (Figure B-9). The command is ***Left Shoulder,* ARMS.**

(1) On the command of execution **ARMS**, execute *Port Arms* in two counts.

(2) On count three, release the grasp of the left hand and (without moving the head) place the carbine on the left shoulder with the right hand (with the sights up), keeping the right elbow down. At the same time, regrasp the carbine with the left hand with the heel of the butt between the first two fingers and with the thumb and forefinger touching. The left forearm is horizontal, and the left upper arm is against the side and on line with the back.

(3) On count four, move the right hand to the right side as in the *Position of Attention.*

Figure B-9. Left Shoulder Arms.

b. *Order Arms* from *Left Shoulder Arms* is a four-count movement (Figure B-10, page B-10). The command is ***Order,*** **ARMS.**

(1) On the command of execution **ARMS,** move the right hand up and across the body and grasp the small of the stock, keeping the right elbow down.

(2) On count two (without moving the head), release the grasp of the left hand and, with the right hand, move the carbine diagonally across the body (sights up) about 4 inches from the waist. At the same time, regrasp the handguard just forward of the slipring with the left hand, and resume *Port Arms.* Counts three and four are executed the same as from *Port Arms.*

Figure B-10. Order Arms from Left Shoulder Arms.

B-8. CHANGING POSITIONS

To change positions with the M4-series carbine use the following procedures:

a. *Right Shoulder Arms* from *Port Arms* is a three-count movement. The command is ***Right Shoulder,*** **ARMS.** On the command of execution **ARMS,** release the grasp of the right hand and regrasp the carbine with the heel of the butt between the first two fingers, with the thumb and forefinger touching. Counts two and three are the same as counts three and four from *Order Arms.* When marching, the command is given as the right foot strikes the marching surface.

b. *Port Arms* from *Right Shoulder Arms* is a two-count movement. The command is ***Port,*** **ARMS.** On the command of execution **ARMS,** execute count one of *Order Arms* from *Right Shoulder Arms.* On the second count, release the grasp of the right hand and regrasp the rifle at the small of the stock and come to *Port Arms.* When marching, the command is given as the right foot strikes the marching surface.

c. *Left Shoulder Arms* from *Port Arms* is a two-count movement. The command is ***Left Shoulder,*** **ARMS.** On the command of execution **ARMS,** execute *Left Shoulder Arms* in the same manner as counts three and four from *Order Arms.* When marching, the command is given as the left foot strikes the marching surface.

d. *Port Arms* from *Left Shoulder Arms* is a two-count movement. The command is ***Port,*** **ARMS.** On the command of execution **ARMS,** execute the first two counts of *Order Arms* from *Left Shoulder Arms.* When marching, the command is given as the left foot strikes the marching surface.

e. *Left Shoulder Arms* from *Right Shoulder Arms* is a four-count movement. The command is ***Left Shoulder,*** **ARMS.** On the command of execution **ARMS,** execute count one the same as executing *Order Arms.* On count two, remove the right hand from the butt of the rifle and regrasp the small of the stock *(Port Arms).* Counts three and four are the same movements as from *Port Arms.* When marching, the command is given as the left foot strikes the marching surface.

f. *Right Shoulder Arms* from *Left Shoulder Arms* is a five-count movement. The command is ***Right Shoulder,*** **ARMS.** On the command of execution **ARMS,** execute *Port Arms* in two counts. Counts three, four, and five are the same as from *Port Arms.* When marching, the command is given as the right foot strikes the marching surface.

g. *Present Arms* from *Right Shoulder Arms* or *Left Shoulder Arms,* while in formation, is executed from the *Halt* only. The command is ***Present,*** **ARMS.** On the command of execution **ARMS,** come to *Port Arms* from either shoulder and then execute *Present Arms* (in one count) from *Port Arms.*

h. To resume *Right (Left) Shoulder Arms* from *Present Arms,* the command is ***Right (Left) Shoulder,*** **ARMS.** On the command of execution **ARMS,** execute *Port Arms* in one count and then execute the counts as prescribed from *Port Arms.*

NOTE: Experienced soldiers should be able to execute the 15-count manual of arms in unison from *Order,* to *Right Shoulder,* to *Left Shoulder,* to *Present,* to *Order Arms*. The command is ***Fifteen-Count Manual,*** **ARMS.**

B-9. SLING ARMS—M4-SERIES CARBINE

These basic procedures apply to executing *Sling Arms* with the M4-series carbine.

- Remain at *Sling Arms* during all rest movements.
- All individual and unit drill movements can be executed at *Sling Arms* except *Stack Arms* and *Fix* and *Unfix Bayonets*, which are executed from *Order Arms*.
- When in formation at *Sling Arms*, execute the *Hand Salute* on the command ***Present,*** **ARMS.**
- Platoon leaders and platoon sergeants carry their carbines at *Sling Arms* during all drills and ceremonies, and from this position they execute *Present Arms* only. They do not execute *Unsling Arms.* This also applies to squad leaders when squads drill as separate units.
- When all members of a unit are carrying their carbines at *Sling Arms*, the platoon leaders and platoon sergeants execute *Present Arms* only. They do not execute *Unsling Arms.*

a. From *Order Arms* with the sling(s) loose, the command for *Sling Arms* is ***Sling,*** **ARMS.** On the command of execution **ARMS,** grasp the sling near the upper sling swivel with the left hand, and release the right hand. Place the right hand and arm between the sling and carbine and place the sling over the right shoulder. Regrasp the sling with the right hand so that the wrist is straight, the right forearm is horizontal, the elbow is tight against the side, and the carbine is vertical. Release the grasp of the left hand and move it sharply to the left side as in the position of attention.

b. To return the rifle to *Order Arms* with the sling tight, the command is ***Adjust,*** **SLINGS.** On command of execution **SLINGS,** remove the rifle from the shoulder. Then, grasp the barrel with the right hand and raise it vertically. With the left hand, place the butt of the carbine on the right hip and cradle it in the crook of the right arm. Use both hands to tighten the sling. Grasp the handguards just above the slipring with the left hand and move the weapon to the *Order Arms* position. At the same, move the right hand to the side and grasp the butt of the weapon. After grasping the butt with the right hand, release the grasp of the left hand and return it sharply to the side as in the *Position of Attention* and return to *Order Arms*.

c. From *Order Arms* with sling(s) tight, the command is ***Sling,*** **ARMS.** On the command of execution **ARMS,** reach across the body with the left hand and grasp the handguards just above the slipring and, with the right hand, place the butt on the right hip, cradle the carbine in the crook of the right arm, and use both hands to adjust the sling. Grasp the sling with the left hand near the upper sling swivel and execute *Sling Arms* (as previously described).

d. If an element is at *Order Arms* with the sling loose and the commander wants the sling to be tightened, he commands ***Adjust*** **SLINGS.** On the command of execution **SLINGS,** tighten the sling and move the carbine to order arms (as previously described).

NOTE: Unless otherwise specified, armed elements of a formation fall in at *Order Arms* with slings loose.

e. To return the carbine to *Order Arms* with the sling loose, the command is ***Unsling,*** **ARMS.** On the command of execution **ARMS,** reach across the body with the left hand and grasp the sling at the right shoulder. Release the grasp of the right hand and remove the carbine from the shoulder. Grasp the barrel with the right hand, release the left hand and regrasp the handguards just above the slipring, release the grasp of the right hand, and move the carbine to the *Order Arms* position (as previously described). Move the carbine to the position of *Order Arms* and regrasp the butt with the right hand. After grasping the butt with the right hand, release the grasp of the left hand and return it sharply to the side as in the *Position of Attention* and return to *Order Arms*.

NOTE: All other manual of arms movements with the M4 slung are the same as with the M16-series rifle when slung. This includes situations when the sling is mounted to the side.

B-10. STACK AND TAKE ARMS (M4-SERIES CARBINE)

The squad members execute *Stack* and *Take Arms* from their positions in line formation (at *Normal Interval*) from *Order Arms*. When in line formation, the squad leader commands ***Count,*** **OFF** and then designates the stack men by numbers (2-5-8).

NOTES:
1. M4-series carbines are not compatible with the M16-series rifles when stacking arms. The two types of weapons must be stacked separately or grounded in a manner that will not damage the sights.
2. When the squad is part of a larger unit, *Stack Arms* may be executed in a column formation (when the formation consists of three or more files and the squads are at *Normal Interval*). Second or third squad is designated as the stack squad.

a. **Prepare Slings.** After the stack men are designated, the squad leader then commands ***Prepare,*** **SLINGS.** On the command of execution **SLINGS,** each stack man (or stack squad) reaches across the body with the left hand and grasp the handguards just above the slipring. With the right hand he places the butt on the right hip and cradles the carbine in the crook of the right arm. Using both hands, he adjusts the sling keeper so that a 2-inch loop is formed from the sling keeper to the upper sling swivel. As soon as the loop is prepared, he returns to *Order Arms.*

b. **Stack Arms.** When all stack men have returned to *Order Arms,* the squad leader commands ***Stack,*** **ARMS** (Figure B-11, page B-15).

(1) On the command of execution **ARMS,** each stack man reaches across the body with the left hand and grasps the handguards just above the slipring. The stackman then grasps the barrel of his carbine with his right hand and places the carbine directly in front of and centered on his body with the sling side to the front. The butt (fully extended) is placed on the marching surface so that the heel of the butt (or side of the butt if the sling is side-mounted) is on line with the toes of his footgear. The stack man bends deeply forward at the waist and grasps his carbine with his left hand at the upper portion of the handguard (keeping the carbine vertical at all times). The first two fingers of the left hand hold the inner part of the loop against the carbine. The stack man reaches across the front of the carbine with his right hand, grasps the outer part of the loop, and holds it open for insertion of other carbines.

(2) On the command of execution **ARMS,** the men to the right and left of the stack man perform the following movements simultaneously:

(a) The man on the stack man's right reaches across the body with the left hand and grasps the handguard. He releases the grasp of the right hand, and regrasps the carbine at the small of the stock. He lowers both arms, with elbows locked holding the rifle in a horizontal position with the muzzle to the left and the magazine well to the front.

(b) The man on the stack man's left reaches across the body with the left hand and grasps the rifle at the small of the stock, releases the right hand, and regrasps the handguard. He then lowers both arms, with elbows locked (holding the carbine in a horizontal position with the muzzle to the right and magazine well to the front).

(3) As soon as the stack man has placed his carbine in position, both men move the foot nearest the stack man halfway *(Half Right* or *Half Left)* toward the stack man. The man on the stack man's left inserts the muzzle of his carbine into the loop to a point about halfway between the flash suppressor and the front sight assembly. He holds his carbine in this position until the man on the stack man's right inserts the muzzle of his carbine in a similar manner and above the other muzzle.

(4) Without moving the feet, both men swing the butt of their carbines out and then down to the marching surface, making the stack tight with the butts on line and about 1 foot from the base line. When the stack has been completed, all three men resume the *Position of Attention.*

(5) Additional carbines are passed to the nearest stack on the right (right or left if stacked in column). The men with additional rifles grasp the handguards with the left hand, releases the grasp of the right hand and regrasps the weapon at the barrel. The men release the grasp of the left hand and raise the carbine vertically with the magazine well to the front, wrist held shoulder high, elbow locked, and right arm extended to the right

front. Throughout the pass, the carbine is held vertical with the magazine well to the front.

(a) The man to the left of the stack man then grasps the carbine at the handguard with his left hand. The man passing the additional carbine then releases the rifle and sharply returns to the *Position of Attention.* The man to the left of the stack man then moves the carbine to the right until it is centered on his body, and he grasps the barrel with his right hand, wrist held shoulder high and elbow locked. He then releases the left hand and sharply returns his left hand to the left side as in the *Position of Attention.* He then moves the carbine to his right front.

(b) The stack man receives the carbine and centers it in the same manner as previously described. The man to the left of the stack man sharply returns to the *Position of Attention* after he releases the carbine. Once the stack man has centered the carbine and grasped the barrel with the right hand, he bends forward at the waist and places the carbine in the stack so that it is secure (without damaging the front sight assembly). If there are two additional carbines, the second carbine is passed in the same manner as the first.

c. **Take Arms.** To *Take Arms,* the command is ***Take,*** **ARMS.** On the command of execution **ARMS,** the men return the additional carbines in the same manner as the carbines were received. The stack man secures the stack and holds the loop in the same manner as for stacking arms. The men on the left and right step toward the stack man in the same manner as when stacking arms. Each man reaches down and regrasps his carbine (one hand at the small of the stock and one at the handguard) and brings it to the horizontal position. The man on the right frees his carbine first and resumes *Order Arms.* The man on the left frees his carbine and resumes *Order Arms.* The stack man cradles his carbine and adjusts the sling and sling keeper to its original position and then resumes *Order Arms.*

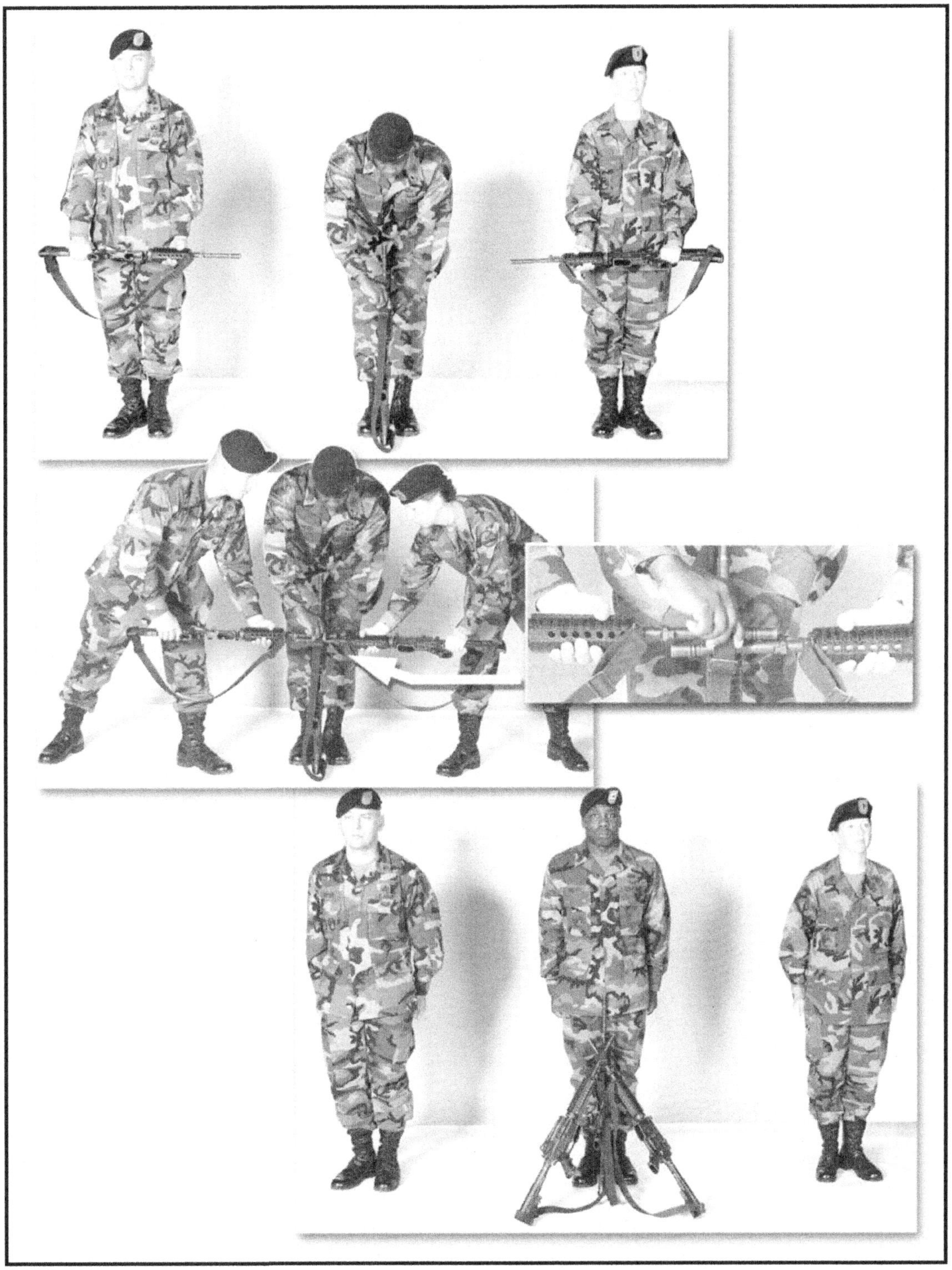

Figure B-11. Stack Arms, M4-series carbine.

This page intentionally left blank.

APPENDIX C
MANUAL OF ARMS—M14 RIFLE

The manual of arms with the M14 (Figure C-1) rifle is executed in the same manner as the manual of arms for the M16-series rifle (Chapter 5) except Inspection Arms, Stack Arms, Order Arms, and some movements of Present Arms.

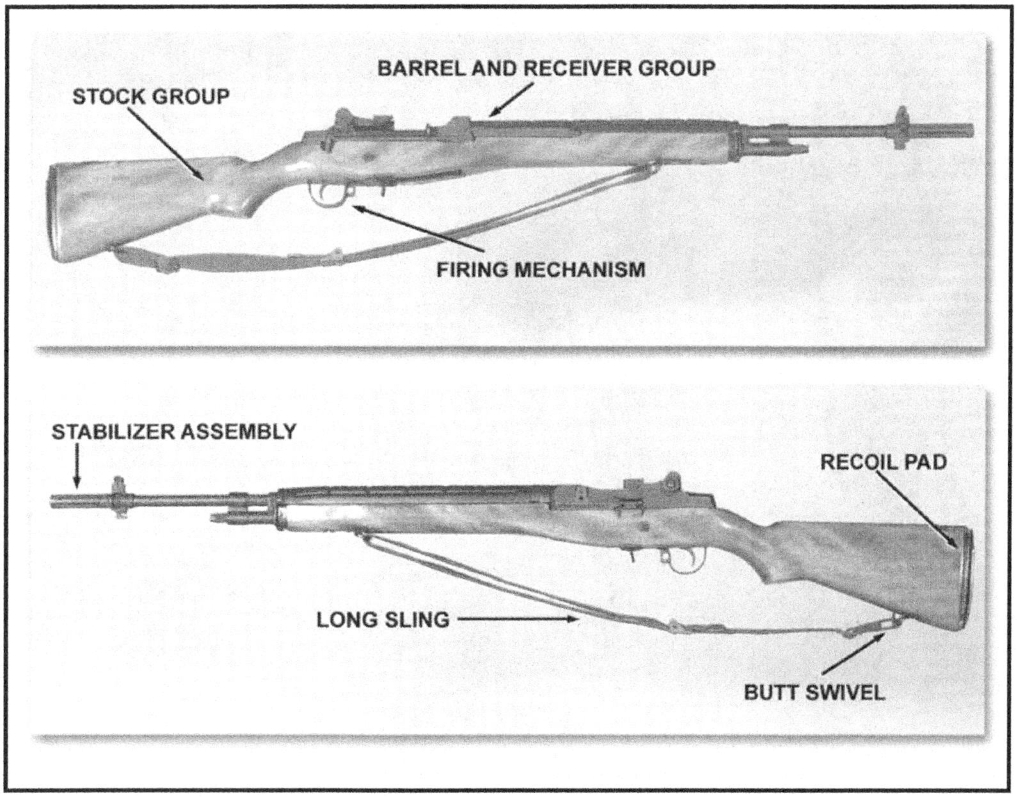

Figure C-1. Nomenclature, M14 rifle.

C-1. ORDER ARMS

Execute *Order Arms* in the same manner as explained for the M16-series rifle except place the butt of the rifle on the marching surface with the toe of the butt on line with the front of the right foot (Figure C-2, page C-2).

C-2. REST POSITION

Execute the *Rest* position in the same manner as explained for the M16-series rifle except keep the toe of the butt on line with the front of the right foot (Figure C-2, page C-2).

C-3. PORT ARMS

Execute *Port Arms* in the same manner as explained for the M16-series rifle except grasp the weapon with the left hand "at the balance." (Figure C-2 shows the location of the balance.)

Figure C-2. Order Arms, Parade Rest, and Port Arms.

C-4. PRESENT ARMS

To execute *Present Arms* with the M14 rifle (Figure C-3), use the following procedures:

a. When at *Order Arms,* the left arm moves across the body on count one with the forearm and wrist straight, fingers and thumb extended and joined, and palm down. The first joint of the forefinger touches the rifle at a point below the bayonet stud. If not in ranks, the head and eyes turn toward the person or Color saluted. On count two, the left hand is moved smartly to the side and the head and eyes are turned to the front.

NOTE: When not at *Order Arms*, execute the movement in the same manner as previously described for the M16-series rifle.

b. When at *Right Shoulder Arms,* the movement is executed by moving the left arm across the chest and touching the first joint of the forefinger to the rear of the receiver.

The left elbow is held so that the left forearm is horizontal. Fingers, thumb, and wrist are held as for *Order Arms*. The palm is down. Count two of the rifle *Salute* at *Right Shoulder Arms* is similar to the return from the rifle *Salute* at *Order Arms*.

c. When at *Left Shoulder Arms,* the *Salute* is executed by moving the right arm across the chest and touching the first joint of the forefinger to the rear of the receiver. The right elbow is held so that the forearm is horizontal. The fingers, thumb, and wrist are as described in paragraph C-4a. The palm is down. Count two is similar to the return from the rifle *Salute* at *Right Shoulder Arms.*

C-5. RIGHT SHOULDER ARMS

Execute *Right Shoulder Arms* in the same manner as explained for the M16-series rifle except change the term "charging" to "receiver" (Figure C-3).

C-6. LEFT SHOULDER ARMS

Execute *Left Shoulder Arms* in the same manner as explained for the M16-series rifle (Figure C-3).

Figure C-3. Present Arms, Right Shoulder Arms, and Left Shoulder Arms.

C-7. INSPECTION ARMS

Inspection Arms with the M14 rifle (Figure C-4) is executed in five counts.

a. Counts one and two are the same as with the M16-series rifle *(Port Arms).* On count three, release the small of the stock and move the right hand forward (fingers extended and joined, thumb on the opposite side of the receiver). Place the knife edge on the hand in contact with the operating rod handle. On count four, press the operating rod handle sharply to the rear and lock the bolt to the rear by pressing the bolt lock with the thumb. At the same time, lower the head and eyes to check the receiver. On count five, raise the head and eyes back to the front and regrasp the small of the stock with the right hand, assuming the *Inspection Arms* position.

b. ***Ready, Port,* ARMS** is the only command that is given from *Inspection Arms.* On the command ***Ready,*** relax the grip of the right hand and hook the top of the forefinger to the rear. On the command ***Port,*** release the operating rod handle (allowing the bolt to go forward) and place the forefinger on the trigger. On the command of execution **ARMS**, pull the trigger and come to *Port Arms.*

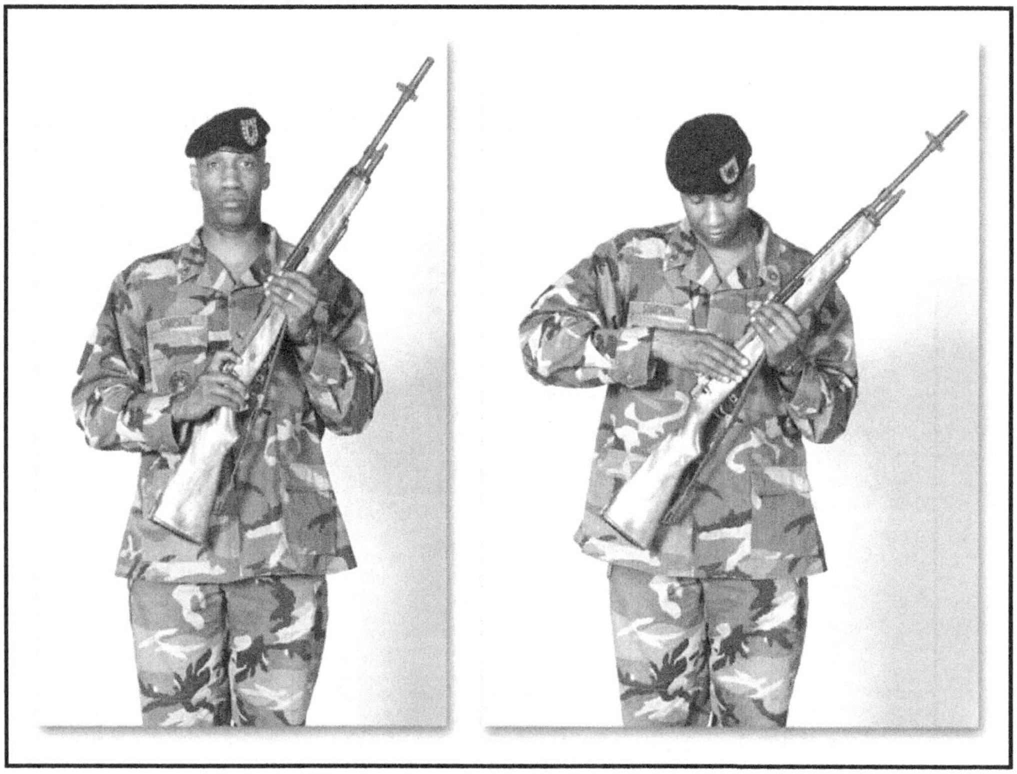

Figure C-4. Inspection Arms.

C-8. FIX AND UNFIX BAYONETS

Execute *Fix* and *Unfix Bayonets* in the same manner as explained for the M16-series rifle.

C-9. SLING ARMS

Execute *Sling Arms* in the same manner as explained for the M16-series rifle.

C-10. STACK AND TAKE ARMS

Execute *Stack Arms* using the following procedures:

a. *Stack Arms* is executed from *Order Arms* only. It may be executed while at *Normal Interval* in a line (squad) or a column formation. *Stack Arms* is not a precise movement; however, it is executed in a military manner.

b. The leader designates the stackmen by numbers (2-5-8) when in a line formation, or by squad (second or third) when in a column formation. After the stackmen have been designated, the leader then commands ***Prepare,*** **SLINGS**.

c. On the command **SLINGS**, each stackman places the butt of his rifle on his right hip and cradles it in the crook of his right arm. He then adjusts the sling keeper to form a 2-inch loop next to the upper sling swivel. As soon as he has prepared the loop, he returns to *Order Arms.*

d. When all stackmen have returned to *Order Arms,* the leader commands ***Stack,*** **ARMS**. On the command **ARMS**, each stackman places his rifle directly in front of and centered on his body with the sling facing to the front. The heel of the rifle butt is on the marching surface on line with the toes of his shoes. He grasps the rifle by the handguard with his left hand. The first two fingers of the left hand hold the inner part of the loop against the rifle. He reaches across the front of the rifle with the right hand, grasps the outer part of the loop, and holds it open for the insertion of the other rifles. He holds the rifle vertical at all times (1, Figure C-5, page C-6).

e. On the command **ARMS**, the men to the left and right of the stackman perform the following movements simultaneously:

(1) The man on the stackman's left raises and centers his rifle in front of his body so that his right hand is shoulder high and the rifle is about 4 inches from his chest, with the sling to the front. He then grasps the rifle with his left hand at the small of the stock and lowers both arms, holding the weapon in a horizontal position (1, Figure C-5, page C-6).

(2) The man on the stackman's right raises and centers his rifle, wrist shoulder high, with the sling facing the front. He then grasps the rifle with his left hand directly below his right hand, and, in the most convenient manner, grasps the rifle with the right hand at the small of the stock. He then lowers both arms, holding the rifle stock. He then lowers both arms, holding the rifle in a horizontal position (1, Figure C-5, page C-6).

(3) As soon as both men have completed these movements, each moves the foot nearest the stackman 18 inches *(Half Left* or *Right)* toward the stackman. In a continuing motion, the man on the stackman's left inserts the muzzle of his rifle into the loop held by the stackman until the bayonet stud protrudes past the far end of the loop (2, Figure C-5, page C-6). He holds his rifle in that position until the man on the stackman's right inserts the muzzle of his rifle through the loop in the same manner and above the muzzle of the rifle of the left man.

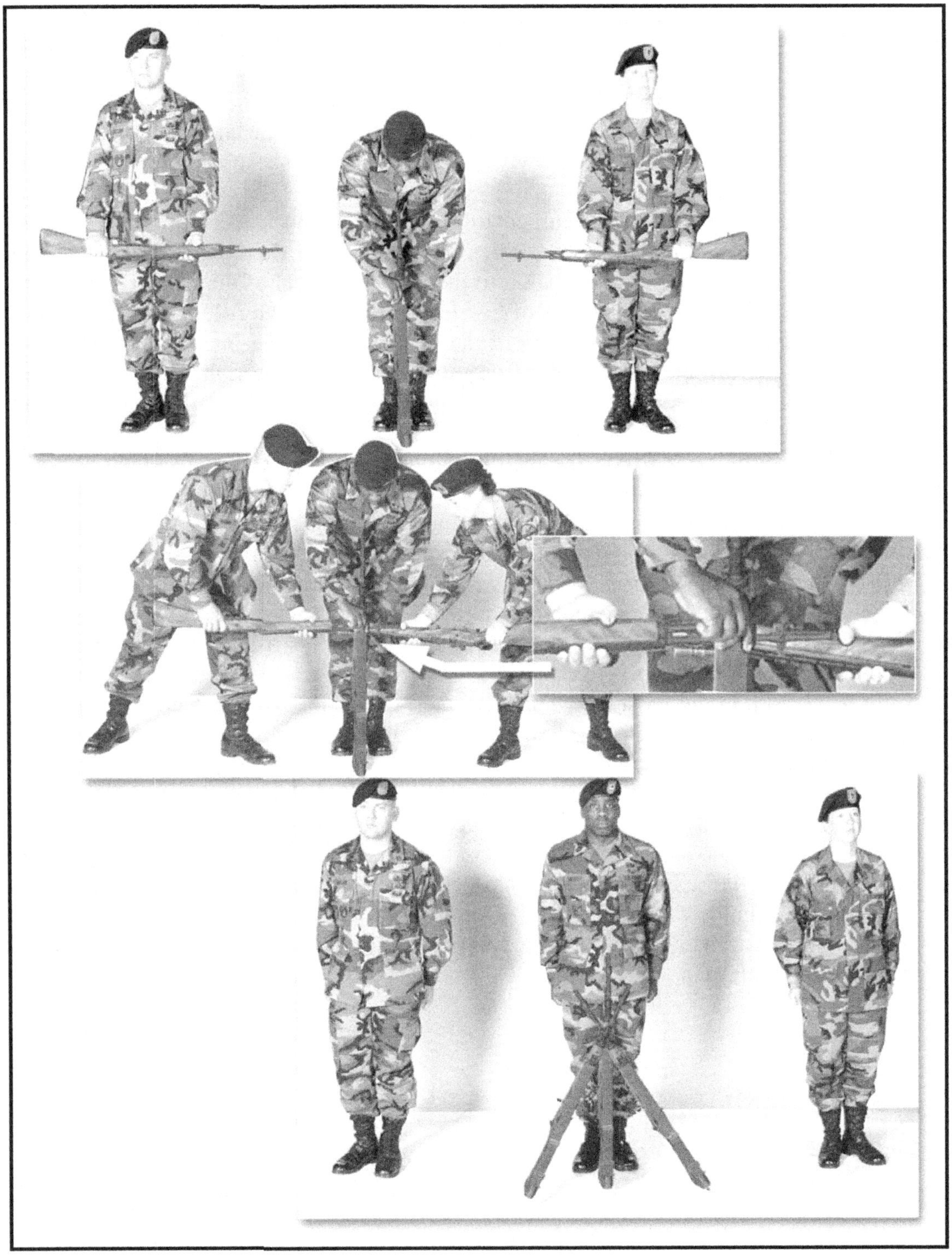

Figure C-5. Stack Arms.

f. When both rifles have been inserted into the loop, the men on each side of the stackman swing the butts of their rifles out and down to the marching surface until the stack is tight with the (two) rifle butts on line and about 2 feet from the baseline.

g. After necessary adjustments have been made, the three men come to *Attention* (3, Figure C-5).

h. Extra rifles are passed to the nearest stack on the right. As each rifle is passed, it is grasped at the upper part of the handguard with the right hand. Then, with the rifle held vertical, it is passed with fully extended arm to the right front. The man on the right grasps the rifle at the balance with his left hand, brings the rifle to the center of his body, and regrasps it at the upper part of the handguard with his right hand. This action continues until the stackman receives the rifle and places it on the stack with his right hand as nearly vertical as possible. He places the rifles on the stack with the slings away from the stack (if the second squad is the stack squad, extra rifles are passed to the left).

i. To *Take Arms,* the leader commands ***Take,*** **ARMS**.

(1) On the command **ARMS**, the stackman passes each extra rifle toward its bearer. The rifles are handled in the manner described for passing them to the stack (left hand at the balance, right hand at the upper part of the handguard). As the men receive their rifles, they come to *Order Arms.*

(2) After all extra rifles have been returned, the stackman grasps the base rifles, holding the loop open as in *Stack Arms.*

(3) The men to the right and left of the stackman take one step to their left and right fronts, respectively. They reach down, grasp their rifles, and bring the rifles to a horizontal position. The man to the right of the stackman frees his rifle from the stack first. Each man returns to *Order Arms* after *Unsling Arms* and retrieves his rifle. He does this by guiding and steadying the rifle with his left hand (as in the next to the last count of *Order Arms).*

(4) The stackman adjusts the sling of his rifle before returning to *Order Arms.*

This page intentionally left blank.

APPENDIX D

MANUAL OF ARMS—M1903 OR M1917 RIFLE

The manual of arms with the M1903 or M1917 rifle (Figure D-1) is executed in the same manner as the manual of arms for the M16-series rifle (Chapter 5) except for minor differences with Order Arms, Port Arms, Left and Right Shoulder Arms, Inspection Arms, and Present Arms. For ease of understanding, the movements are fully described.

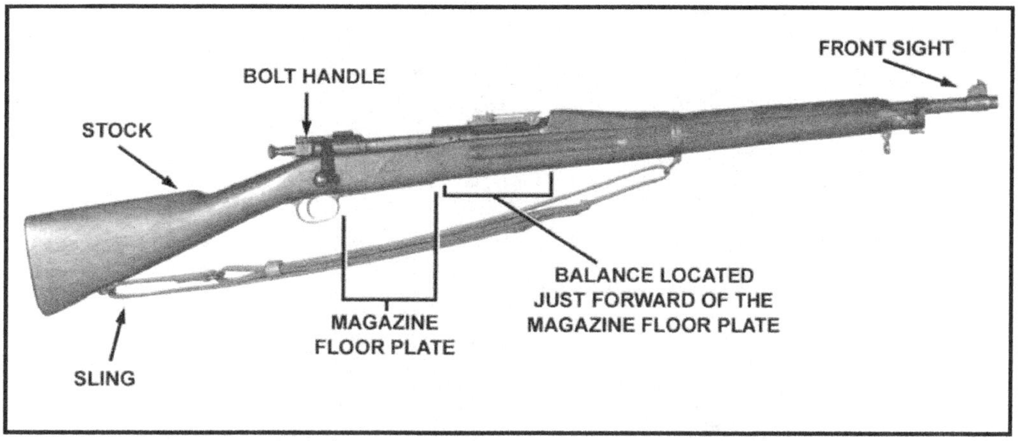

Figure D-1. Nomenclature, M1903 or M1917 rifle.

NOTE: The term "at the balance" refers to where the trailing edge of the top handguard meets the front edge of the breech. The trailing edge of the hand is placed in this area. Figure D-1 shows the location of the balance.

D-1. ORDER ARMS

Execute *Order Arms* using the following procedures:

a. Assume *Order Arms* on the command **FALL IN** or from *Parade Rest* on the command of execution **ATTENTION**.

b. At *Order Arms*, maintain the *Position of Attention* with the rifle. Place the butt of the rifle on the marching surface with sights to the rear and touching the right foot. Place the toe of the butt on line with the front of the right foot. Secure the rifle with the right hand in a "U" formed by the fingers (extended and joined) and thumb. Keep the right arm as in the *Position of Attention* and hold the rifle with the right thumb and fingers with the fingers pointed generally downward and touching the sides of the handguards. Keep the right hand and arm behind the rifle (Figure D-2).

Figure D-2. Position of Attention and Parade Rest.

D-2. REST POSITION

The rifle *Rest* positions are commanded and executed the same as individual drill with the following additions:

a. On the command of execution **REST** of ***Parade,*** **REST**, thrust the muzzle forward keeping the toe of the butt on line with the front of the right foot and the right arm straight (Figure D-2).

b. Execute *Stand at Ease* with the rifle in the same manner as *Parade Rest* except turn the head and eyes toward the commander.

c. On the command **AT EASE** or **REST**, keep the butt of the rifle in place as in *Parade Rest*.

D-3. PORT ARMS

Execute *Port Arms* from *Order Arms* using the following procedures;

a. *Port Arms* from *Order Arms* is a two-count movement. The command is ***Port,*** **ARMS**. On the command of execution **ARMS**, grasp the rifle with the right hand and raise the rifle diagonally across the body, keeping the right elbow down (without strain). With the left hand, simultaneously grasp the rifle at the balance so that the rifle is about 4 inches from the waist. On count two, regrasp the rifle at the small of the stock with the right hand. Hold the rifle diagonally across the body, about 4 inches from the waist, the right forearm horizontal, and the elbows close to the sides (Figure D-3).

b. *Order Arms* from *Port Arms* is executed in three counts. The command is ***Order,* ARMS**. On the command of execution **ARMS**, move the right hand up and across the body and firmly grasp the rifle just forward of the lower band without moving the rifle, and keep the right elbow down without strain. On count two, move the left hand from the balance and lower the rifle to the right side until it is about 1 inch from the marching surface. Guide the rifle to the side by placing the forefinger of the left hand at the forward edge of the upper band, fingers and thumb extended and joined, palm to the rear. On count three, move the left hand sharply to the left side, gently lower the rifle to the marching surface, and resume the position of *Order Arms* (Figure D-3).

Figure D-3. Order Arms from Port Arms.

D-4. PRESENT ARMS

Execute *Present Arms* using the following procedures:

a. *Present Arms* from *Order Arms* is a three-count movement. The command is ***Present,* ARMS**. On the command of execution **ARMS**, execute *Port Arms* in two counts. On count three, twist the rifle with the right hand so that the sights are to the rear, and move the rifle to a vertical position about 4 inches in front of and centered on the body. Lower the rifle until the left forearm is horizontal; keep the elbows in at the sides. Keep the left thumb alongside and touching the handguard.

NOTE: Incline the barrel slightly backward to ensure that the weapon is vertical.

b. *Order Arms* from *Present Arms* is a four-count movement. The command is ***Order,*** **ARMS**. On the command of execution **ARMS,** return the rifle to *Port Arms*. Counts two, three, and four are the same as *Order Arms* from *Port Arms* (Figure D-4).

c. *Port Arms* is assumed en route to or from *Present Arms* when going to or from *Right Shoulder* or *Left Shoulder Arms*. *Present Arms* from or to *Port Arms* is a one-count movement.

d. When rendering reports or courtesy to an individual from *Order Arms*, execute *Present Arms* and turn the head and eyes toward the individual addressed. *Order Arms* is executed automatically upon acknowledgment of the *Salute*.

Figure D-4. Order Arms from Present Arms.

D-5. RIGHT SHOULDER ARMS

Execute *Right Shoulder Arms* using the following procedures:

a. *Right Shoulder Arms* from *Order Arms* is a four-count movement. The command is ***Right Shoulder,*** **ARMS**. Execute count one, the command of execution **ARMS,** the same as in executing *Port Arms*. On count two, release the grasp of the right hand and grasp the heel of the butt between the first two fingers with the thumb and forefinger touching. On count three (without moving the head), release the grasp of the left hand (without changing the grasp of the right hand), twist the rifle so that the sights are up, and place the weapon onto the right shoulder, moving the left hand to the small of the stock to guide the rifle to the shoulder. Keep the fingers and thumb (left hand) extended and joined with the palm turned toward the body. The first joint of the left forefinger touches the rear of the cocking piece. Keep the left elbow down, and keep the right forearm horizontal with the right upper arm against the side and on line with the back. On count four, sharply move the left hand back to the left side as in the *Position of Attention* (Figure D-5).

b. *Order Arms* from *Right Shoulder Arms* is a four-count movement. The command is ***Order,*** **ARMS.** On the command of execution **ARMS**, without moving the head and without changing the grasp of the right hand, press down quickly and firmly on the butt of the rifle with the right hand and twist the weapon (with the sights up), guiding it diagonally across the body and about 4 inches from the waist. Grasp the rifle with the left hand at the balance. On count two, move the right hand up and across the body and firmly grasp the rifle just forward of the lower band without moving the rifle; keep the right elbow down without strain. Counts three and four are the same as from *Port Arms* to *Order Arms* (Figure D-5, page D-6).

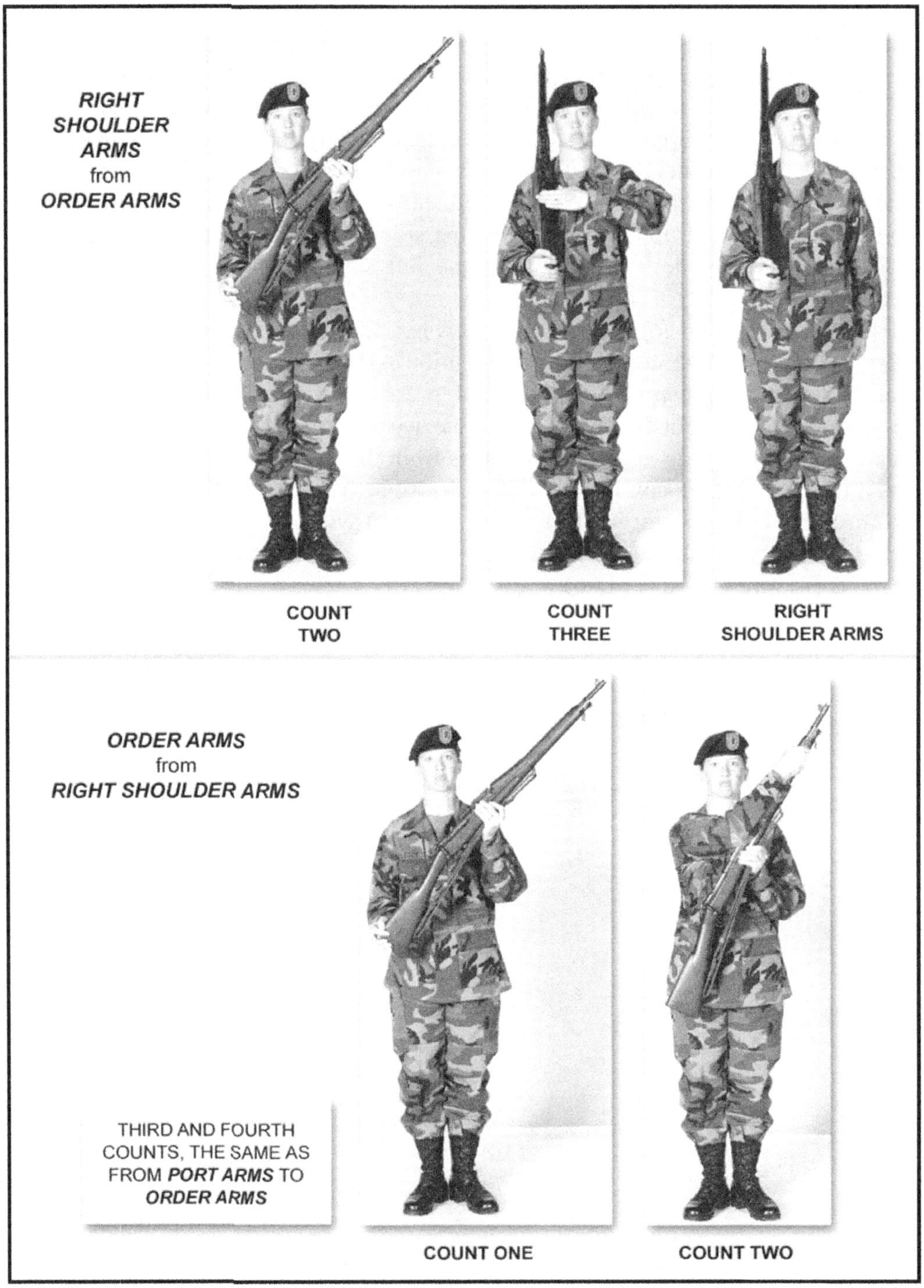

Figure D-5. Right Shoulder Arms from Order Arms and Order Arms from Right Shoulder Arms.

D-6. LEFT SHOULDER ARMS

Execute *Left Shoulder Arms* using the following procedures:

a. *Left Shoulder Arms* from *Order Arms* is a four-count movement. The command is ***Left Shoulder,* ARMS**. On the command of execution **ARMS**, execute *Port Arms* in two counts. On count three, release the grasp of the left hand and (without moving the head) place the rifle on the left shoulder with the right hand (with the sights up), keeping the right elbow down. At the same time, regrasp the rifle with the left hand with the heel of the butt between the first two fingers and with the thumb and forefinger touching. The left forearm is horizontal, and the left upper arm is against the side and on line with the back. On count four, move the right hand to the right side as in the *Position of Attention* (Figure D-6).

b. *Order Arms* from *Left Shoulder Arms* is a five-count movement. The command is ***Order,* ARMS**. On the command of execution **ARMS,** move the right hand up and across the body and grasp the small of the stock, keeping the right elbow down. On count two (without moving the head), release the grasp of the left hand and with the right hand move the rifle diagonally across the body (sights up) about 4 inches from the waist. At the same time, regrasp the rifle at the balance with the left hand, and resume *Port Arms* (Figure D-6). Counts three, four, and five are the same as *Order Arms* from *Port Arms*.

Figure D-6. Left Shoulder Arms from Order Arms and Order Arms from Left Shoulder Arms.

D-7. INSPECTION ARMS

Execute *Inspection Arms* using the following procedures:

a. *Inspection Arms* from *Order Arms* is a five-count movement. The command is ***Inspection,*** **ARMS.** On the command of execution **ARMS,** execute *Port Arms* in two counts. On count three, grasp the bolt handle with the thumb and forefinger of the right hand, rotate the handle upward and draw the bolt back. On count four, lower the head and eyes and visually inspect the magazine. On count five, return the head and eyes to the front as in the *Position of Attention*, and come to *Inspection Arms* (Figure D-7).

NOTE: The method for inspecting the rifle by an inspecting officer is explained in Chapter 7.

b. ***Port,*** **ARMS** is the only command given from *Inspection Arms*. On the command ***Port,*** push the bolt forward and rotate the handle down locking the bolt in place, then release the grasp of the right hand and regrasp the small of the stock with the right forefinger on the trigger. On the command **ARMS,** pull the trigger and resume *Port Arms* (Figure D-7).

Figure D-7. Inspection Arms from Order Arms and Port Arms from Inspection Arms.

APPENDIX E

MANUAL OF ARMS—SPECIALITY WEAPONS

This appendix contains procedures for executing the manual of arms with the M249 squad automatic weapon (Figure E-1), M203 grenade launcher, shotgun, and pistol.

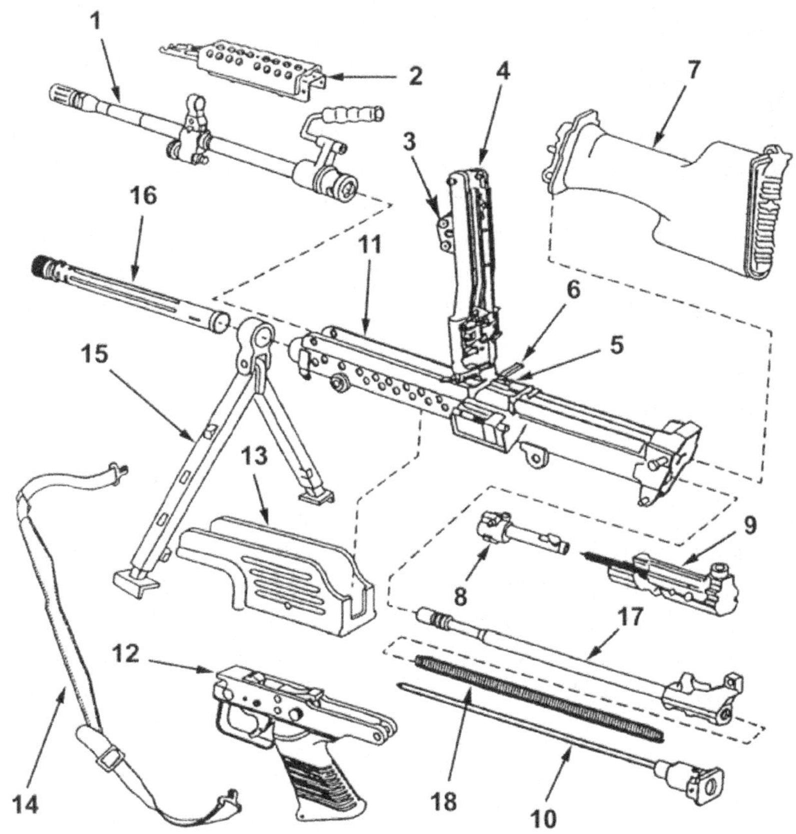

(1) BARREL ASSEMBLY	(10) RETURN ROD AND TRANSFER MECHANIZM ASSEMBLY
(2) HEAT SHIELD ASSEMBLY	(11) RECEIVER ASSEMBLY
(3) REAR SIGHT ASSEMBLY	(12) TRIGGER ASSEMBLY
(4) COVER AND FEED MECHANISM ASSEMBLY	(13) HANDGUARD ASSEMBLY
(5) FEED TRAY ASSEMBLY	(14) SLING AND SNAP HOOK ASSEMBLY
(6) COCKING HANDLE ASSEMBLY	(15) BIPOD
(7) BUTTSTOCK AND BUFFER ASSEMBLY	(16) GAS CYLINDER ASSEMBLY
(8) BOLT ASSEMBLY	(17) PISTON ASSEMBLY
(9) SLIDE ASSEMBLY	(18) RETURN SPRING

Figure E-1. Nomenclature, M249.

E-1. M249 SQUAD AUTOMATIC WEAPON

The M249 is carried at *Sling Arms* in the same manner as the M16. **All** individual drill movements (as applicable) are executed while at *Sling Arms*. When *At Ease* or *Rest* is commanded (from *Parade Rest),* the soldier unslings the weapon and places the butt of the weapon on the marching surface beside his right foot. The only manual of arms movements that may be executed are *Present Arms (Present Arms* at *Sling Arms), Unsling Arms, Inspection Arms,* and *Ready Port Arms.* (*Inspection Arms* and *Ready Port Arms* may only be executed from the *Order Arms* position.) During all other manual of arms movements, the soldier remains at *Sling Arms*.

a. **Inspection Arms with the M249.** The command for this movement is ***Inspection,*** **ARMS**.

(1) On the command of execution **ARMS**, execute *Port Arms* from *Order Arms* in two counts. On count three, grasp the pistol grip with the right hand and reach under the weapon with the left hand and grasp the cocking handle (the back of the hand faces toward the muzzle). On count four, sharply push the cocking handle rearward so that the bolt is locked to the rear. On count five, sharply return the cocking handle to its original position. On count six, bring the left hand between the body and weapon, unlock the feed tray cover (with the thumb and forefinger), and raise the feed tray cover. On count seven, grasp the feed tray with the left hand, open the feed tray and feed tray cover as far as possible, and observe the chamber. On count eight, regrasp the handguard with the left hand and the comb of the stock with the right hand, and come to *Inspection Arms* (Figure E-2).

(2) ***Ready, Port*** **ARMS** is the only command that is given from *Inspection Arms.* On the command ***Ready***, grasp the pistol grip with the right hand and the feed tray cover with the left hand and close the feed tray cover. On the command ***Port***, grasp the cocking handle with the left hand and sharply push the cocking handle to its most rearward position. On the command **ARMS**, pull the trigger with the right forefinger and ride the bolt forward with the left hand, then return to the *Port Arms* position (Figure E-2).

Figure E-2. Inspection Arms with the M249.

Figure E-2. Inspection Arms with the M249 (continued).

E-2. M203 GRENADE LAUNCHER

This paragraph contains the procedures for executing the manual of arms movements with the M203 weapon.

a. **Basic Procedures.** The following basic procedures apply to the M203 weapon.

(1)The M203 weapon is carried at *Sling Arms.* All individual drill movements are executed as prescribed while remaining at *Sling Arms.*

(2) The only manual of arms movements that may be executed are *Present Arms (Hand Salute), Port Arms,* and *Inspection Arms.* Individuals armed with the M203 weapon remain at *Sling Arms* during other manual of arms movements; or, they return to *Sling Arms* from *Port Arms* when other members with M16-series rifles are commanded to other positions.

b. **Manual Movements with the M203.** To execute manual movements with the M203 (Figure E-3), use the following procedures:

(1) ***Present Arms.*** The *Hand Salute* with the M203 is executed in the same manner as prescribed for the M16-series rifle (while at *Sling Arms).*

SLING ARMS PORT ARMS INSPECTION ARMS

Figure E-3. Manual movements with the M203.

(2) ***Port Arms.*** Execute *Port Arms* (Figure E-3) in the same manner as with the M16-series rifle from *Sling Arms* except grasp the weapon at the bottom of the hand guard on the barrel of the M203.

(3) ***Inspection Arms.*** Execute *Inspection Arms* (Figure E-3) with the M203 weapon the same as *Inspection Arms* (while at *Sling Arms*) with the M16-series rifle. After

completing the movements for *Inspection Arms,* press the barrel release latch with the left thumb, slide the barrel to the left, and visually inspect the chamber.

(4) ***Ready Port Arms.*** Execute *Ready Port Arms* the same as *Ready Port Arms* with the M16-series rifle with one exception: on the command ***Ready,*** the first action is to close the barrel and pull the trigger of the grenade launcher.

E-3. SHOTGUN (RIOT)

Except for *Inspection Arms,* execute the manual of arms with the shotgun in the same manner as prescribed for the M16-series rifle. Carry the weapon at the *Ready Position* when physically guarding prisoners (Figure E-4). The courtesy (*Salute*) is not rendered to individuals while performing as a prisoner guard; however, during the playing of the National Anthem, "To the Color," "Ruffles and Flourishes," "Reveille," or "Taps," command the prisoners to *Attention*, remain facing them, and execute *Present Arms* from the *Ready Position*.

Figure E-4. Ready Position.

a. **Ready Position.** Assume the *Ready Position* in the most convenient manner. Place the butt of the weapon on the right hip and grasp the weapon with the right hand forward of the trigger guard. Bend the right arm comfortably, and keep the gun barrel at an angle from the body and the muzzle skyward. To ease fatigue, carry the weapon in the same manner on the left hip.

b. **Inspection Arms With the Shotgun.** *Inspection Arms* from *Order Arms* is a three-count movement. Execute *Inspection Arms* as follows:

(1) The command is ***Inspection,*** **ARMS.** On the command of execution **ARMS,** execute *Port Arms* (M16-series) in two counts. On count three, push the slide to the right with the left hand (over the sling). Twist the weapon to visually inspect the receiver through the ejection port and then come to *Inspection Arms.*

(2) ***Ready, Port,*** **ARMS** is the only command that may be given from *Inspection Arms*. On the command ***Ready,*** take no action. On the command ***Port,*** move the slide to the left (close the bolt) and place the right forefinger on the trigger. On the command **ARMS,** pull the trigger and come to *Port Arms.*

E-4. AUTOMATIC PISTOLS

This paragraph contains the procedures for executing the manual of arms with automatic pistols (FM 23-35). All individual drill movements are executed as prescribed with the weapon secure in the holster. The only manual of arms movements that may be executed are *Present Arms (Hand Salute)* and *Inspection Arms*. During other manual of arms movements, remain at attention. The position of *Raised Pistol* represents the position of *Port Arms* when challenging as a sentinel on guard duty.

a. **Present Arms.** The *Hand Salute* is executed in the same manner as prescribed in individual drill for the M16-series rifle.

b. **Raised Pistol.** For instructional purposes, the command ***Raised,*** **PISTOL** is used to teach individuals the procedures for challenging.

(1) On the command of execution **PISTOL,** unbutton the flap of the holster with the right hand and grasp the pistol grip. Withdraw the pistol from the holster, raising the right forearm to the front until the hand is about 6 inches in front of the right shoulder, and the right elbow is against the right side. Hold the pistol grip in a "U" formed by the thumb and last three fingers. Extend the forefinger outside and alongside the trigger guard. Point the muzzle outward and up at about a 30-degree angle (Figure E-5, page E-8).

(2) On the command ***Return,*** **PISTOL,** return the weapon in the most convenient manner to the holster, fasten the flap, and come to attention.

c. **Inspection Arms.** The command for this movement is ***Inspection,*** **ARMS.**

(1) On the command of execution **ARMS,** execute raised pistol and then, without lowering the right hand, reach across the body and grasp the slide with the left thumb and the first two fingers so that the thumb is on the left side of the slide. Shift the grip of the right hand so that the right thumb engages the slide stop. Push the slide fully to the rear and engage the stop in its notch with the right thumb. Return the left hand to the left side and remain at raised pistol.

(2) ***Ready, Port,*** **ARMS** is the only command that may be given from inspection arms. On the command ***Ready,*** place the right thumb on the slide stop. On the command ***Port,*** press down on the slide stop, allowing the slide to go forward, and then place the forefinger on the trigger. On the command **ARMS,** pull the trigger and return the weapon to the holster, fasten the flap, and come to *Attention.*

(3) To execute *Inspection Arms* with the magazine in the weapon, execute *Raised* pistol and then remove the magazine and place it between the belt and clothing before locking the slide to the rear. After pulling the trigger, and before returning the pistol to the holster, return the magazine to the weapon.

(4) If the pistol is inspected in ranks and the inspecting officer takes the weapon, lower the right hand smartly to the right side. When the inspecting officer is ready to return the pistol, raise the right hand back to the raised position and accept the weapon. After the pistol is returned, allow the slide to go forward, pull the trigger, return the pistol to the holster, and come to *Attention.*

RAISED PISTOL   INSPECTION ARMS

Figure E-5. Manual of arms, automatic pistols.

APPENDIX F

MANUAL OF ARMS—SABER AND SWORD

The saber is worn by officers while participating in ceremonies with troops under arms, or as directed. It is carried on the left side of the body attached to the belt by the scabbard chain with the guard of the saber to the rear. The sword is worn by all platoon sergeants and first sergeants while participating in ceremonies with troops under arms, or as directed. It is carried in the same manner as the officer's saber.

F-1. NOMENCLATURE

The nomenclature for the saber is saber for all officers, model 1902. The blade is 31 inches long. The nomenclature for the sword is noncommissioned officer's sword, model 1840. Figure F-1 shows the nomenclature for pertinent parts of the saber (sword) and scabbard.

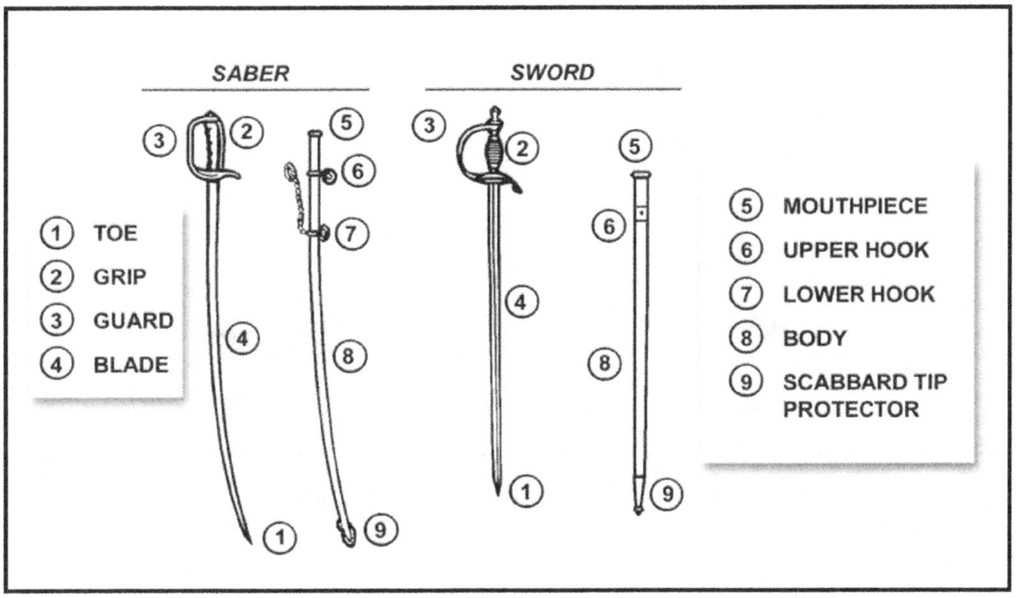

Figure F-1. Nomenclature, saber and sword.

F-2. STANDING MANUAL OF ARMS

Execute *Standing* with the saber (sword) using the following procedures:

a. **Attention.** This is the position before the command ***Draw*, SABER (SWORD)** and after the command ***Return, SABER* (SWORD).** The hands are behind the trouser seams and the thumbs touch the first joint of the forefingers (Figure F-2, page F-2).

Figure F-2. Position of Attention.

b. **Draw Saber (Sword).** Figure F-3 shows the sequence for executing *Draw Saber (Sword).*

Figure F-3. Draw Saber (Sword).

(1) At the preparatory command **Draw,** grasp the scabbard with the left hand turning the scabbard clockwise 180 degrees, tilting it forward to form an angle of 45 degrees with the ground. Take the saber (sword) grip in the right hand and pull the saber about 6 inches from the scabbard. The right forearm should now be roughly parallel to the ground.

(2) On the command of execution **SABER (SWORD),** the saber (sword) is pulled out of the scabbard and held in the position of *Carry Saber (Sword).* The saber (sword) should be held with the inner blade-edge riding in a vertical position along the forward tip of the right shoulder (Figure F-4, page F-5).

c. **Carry Saber (Sword).** The *Carry Saber (Sword)* position is assumed under the following situations:

- To give commands.
- To change positions.
- By officers when officially addressing (or when officially addressed by) another officer, if the saber is drawn.
- By NCOs when officially addressing a soldier, or when officially addressed by an officer, if the sword is drawn.
- Before returning the saber (sword) to the scabbard.
- At the preparatory command for (and while marching at) quick time.
- (1) At *Carry Saber (Sword)* (Figure F-4), the officer (NCO) is at the *Position of Attention.* The saber (sword) is held in the right hand; the wrist is as straight as possible with the thumb along the seam of the trouser leg. The point of the blade rests inside the point of the shoulder and not along the arm. The saber (sword) is held in this position by the thumb and forefinger grasping the grip, and it is steadied with the second finger behind the grip.

(2) *Present Saber (Sword)* may be executed from the *carry* when serving in the capacity of commander of troops or serving in a command that is not part of a larger unit. On the preparatory command of ***Present,*** the saber (sword) is brought to a position (at the rate, of two counts) approximately four inches from the nose so that the tip of the saber (sword) is six inches from the vertical (1, Figure F-5, page F-6). At the command of execution **ARMS,** the right hand is lowered (at the rate of two counts) with the flat of the blade upward, the thumb extended on the left side of the grip (2, Figure F-5, page F-6), and the tip of the saber (sword) about six inches from the marching surface.

(3) On the command ***Order*** **ARMS,** the saber (sword) is returned to the position of *Carry Saber (Sword).*

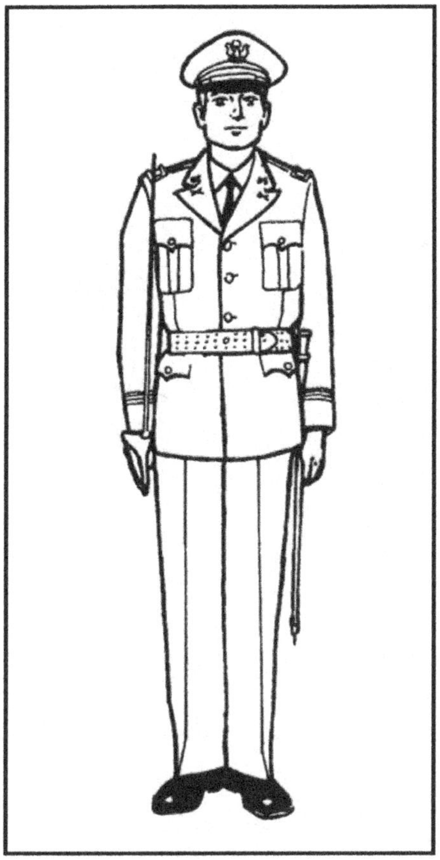

Figure F-4. Carry Saber (Sword).

Figure F-5. Present Saber (Sword).

d. **Parade Rest.** This position is assumed without moving the saber (sword) from the *Order Arms* position. At the command of execution, the left foot is moved about 10 inches to the left (of the right foot), and the left hand is placed in the small of the back, fingers extended and joined, palm to the rear (Figure F-6). At the command of execution **ATTENTION,** the left hand and foot are returned to the *Position of Attention.*

Figure F-6. Parade Rest.

NOTE: Whenever the saber (sword) is at the *Order Arms* position the saber (sword) is straight, not at an angle inward or outward in relationship to the body.

e. **Return Saber (Sword).** This movement is executed from *Carry Saber (Sword)* in three counts.

(1) At the preparatory command ***Return*** of the command ***Officers (Noncommissioned officers), Return,*** **SABER (SWORD),** the saber (sword) is brought to a vertical position (1, Figure F-7, page F-8). The forearm (wrist) is held parallel to the marching surface about three inches from the body; the guard is pointed to the left.

(2) At the command of execution **SABER (SWORD),** three actions take place simultaneously: the saber (sword) is pivoted downward toward the guard, at the same time grasp the scabbard with the left hand just above the upper brass ring mounting. Tilt it forward and turn it clockwise 180 degrees. The scabbard should form a 45-degree angle with the ground, and the saber (sword) bearer turns his head to the left and, looks down to observe the mouthpiece of the scabbard (the shoulders remain squared to the front and level). As smoothly and as quickly as possible, the saber (sword) is inserted into the scabbard and stopped so that about 12 inches of the blade is showing; the right forearm

(wrist) is horizontal to the marching surface and three inches from the body (2, Figure F-7).

(3) At the command of execution **CUT** of the command ***Ready,* CUT**, the saber (sword) is thrust smartly into the scabbard, the scabbard is rotated so that its tip is forward, and the saber (sword) bearer comes to *Attention* (3, Figure F-7).

Figure F-7. Return Saber (Sword).

K-3. MARCHING MANUAL OF ARMS

While marching, the saber (sword) is carried with the inner blade edge riding in a vertical position along the forward tip of the right shoulder (Figure F-8).

Figure F-8. Marching position.

a. **Eyes Right While Marching.** The command ***Eyes,* RIGHT** is executed while marching at *Carry Saber (Sword).* The command ***Ready*** is given as the right foot strikes the marching surface (no action is taken). The second time the right foot strikes the marching surface, the command ***Eyes*** is given and the saber (sword) is brought to the position (count one position) of *Present Arms.* No action is taken the third time the right foot strikes the marching surface. The fourth time the right foot strikes the marching surface, the command ***Right*** is given. As the foot strikes the marching surface, the head is turned sharply to the right at a 45 degree angle, and the saber (sword) is brought downward (Figure F-9, page F-10).

NOTE: The initial preparatory command ***Ready*** is only used when marching with a guidon bearer.

Figure F-9. Eyes right while marching.

(1) While marching at *Present Saber (Sword),* the right arm is swung naturally (nine inches to the front and six inches to the rear) in a vertical plane, flexing the wrist to keep the tip of the blade level (about six inches) above the marching surface. This requires extending the wrist on the forward movement and elevating the wrist on the rearward movement.

(2) The command ***Ready,*** **FRONT** is executed as follows: The command ***Ready*** *is* given as the right foot strikes the marching surface (no action is taken). The second time the right foot strikes the marching surface, a second command ***Ready*** *is* given (again, no action is taken). The saber (sword) is returned to the *Order* position while maintaining the arm swing (subparagraph b) as the right foot strikes the marching surface the third time. The fourth time the right foot strikes the marching surface the command **FRONT** is given. The head is turned sharply to the front as the right foot strikes the marching surface and the saber (sword) is returned to the *Carry* position the next time the left foot strikes the marching surface.

b. **Port Arms.** *Port Arms* (Figure F-10) is executed on the preparatory command ***Double Time*** of the command ***Double Time,*** **MARCH.** This position is assumed only from the position of *Carry Saber (Sword).* The right arm swings naturally across and 6 inches in front of the body. The saber (sword) is canted 45 degrees from the vertical with the guard pointed to the left. The left hand grasps the scabbard.

Figure F-10. Port Arms.

This page intentionally left blank.

APPENDIX G

SYMBOLS

This appendix contains an explanation of the symbols used in the figures in this manual.

BATTALION COMMANDER

COMMANDER OF TROOPS

COMPANY COMMANDER

EXECUTIVE OFFICER (COMPANY)

PLATOON LEADER

COMMAND SERGEANT MAJOR

STAFF OF UNIT COMMANDER

BAND MASTER

DRUM MAJOR

FIRST SERGEANT

PLATOON SERGEANT

PLATOON SERGEANT (ALTERNATE POSITION)*

SQUAD LEADER

SECTION LEADER

FIRE TEAM LEADER

SQUAD MEMBER

PERSONAL FLAG (IF APPROPRIATE)

GUIDON OR ORGANIZATIONAL COLOR(S) BEARER

COLOR OR FLAG BEARER

COLOR GUARD

COLOR GUARD W/ NATIONALS

COLOR GUARD W/O NATIONALS

FLAG POLE

SALUTE BATTERY

PARADE GROUND MARKER

DIRECTION TROOPS ARE FACING OR MARCHING

NEW DIRECTION

CLERGY

* NOTE: ALL DASH LINE SYMBOLS INDICATE ALTERNATE POSITIONS.

This page intentionally left blank.

APPENDIX H
MANUAL OF THE GUIDON

The guidon is a company, battery, or troop identification flag. It is present at all unit formations unless otherwise directed by the commander.

When armed with a rifle, the guidon bearer slings the weapon behind his back with the sling diagonally across his chest and the muzzle end up and to the right.

H-1. BASIC GUIDON POSITIONS

This paragraph describes basic guidon positions and how to assume them.

a. When the unit is formed, the guidon bearer is one step in front of and two 15-inch left steps to the right of and facing the person forming the unit (the person forming the unit is facing the unit.) If the first sergeant forms the unit, the guidon bearer steps forward three steps on the command **POST**.

b. To face with the guidon, raise the ferrule 1 inch off the marching surface by bending the right elbow slightly. After executing the movement, automatically lower the ferrule to the *Order* position. Keep the staff vertical throughout the movement.

c. Execute facing movements, marching movements, *Present Arms (Present Guidon)*, and *Rest* movements with the unit. During manual of arms movements, remain at *Order Guidon* except for *Present Arms* (Figure H-1, page H-2).

d. When in a line formation and the company is faced to the right for a marching movement, face to the right in marching, assume the *Double-Time* position (Figure H-7, page H-8) and double-time from that position to a position five steps forward of and centered on the squad leaders of the first platoon. If the company commander joins the formation at the head of the column (six steps in front of and centered on the squad leaders), execute two left steps.

e. If a **platoon** is authorized a guidon or phase banner (in training units for example), the bearer's post is one step in front of and two 15-inch steps to the right of and facing the person forming the platoon. When the formation is faced to the right for a marching movement, the bearer executes in the same manner as explained above, except that his post is three steps in front of and centered on the squad leaders. If the platoon leader is present and at his post, the bearer's post is one step to the rear and two 15-inch steps to the left of the platoon leader.

f. When the guidon bearer is marching at his post in a column formation during a *Column Left (Right)*, he executes a *Column Half Left (Right)* on the command of execution and inclines as necessary to maintain his post.

NOTE: Execute raised guidon only on the preparatory commands for ***Present,*** **ARMS*;*** ***Eyes,*** **RIGHT**; and on the preparatory command ***Order*** of ***Order,*** **ARMS** following the execution of *Present Arms* and *Eyes Right*. Also execute *Raised Guidon* on the preparatory command ***Ready*** of ***Ready,*** **FRONT.**

H-2. ORDER GUIDON

At *Order Guidon* (*Position of Attention),* keep the ferrule on the marching surface and touching the outside of the right foot, opposite the ball of the right foot. Hold the staff in the right hand in the "U" formed by the fingers (extended and joined pointing downward) and thumb. Keep the right hand and arm behind the staff. Rest the staff against the hollow of the shoulder (Figure H-1).

Figure H-1. Order Guidon.

H-3. REST POSITIONS

To assume *Rest* positions, execute the following actions.

a. On the preparatory command ***Parade,*** slide the right hand up the staff until the forearm is horizontal and grasp the staff (1, Figure H-2). On the command of execution **REST**, thrust the staff straight forward keeping the ferrule on the marching surface until the arm is fully extended, and at the same time, execute *Parade Rest* as in individual drill (2, Figure H-2).

b. *Stand At Ease* is the same as *Parade Rest,* except that the eyes and head are turned toward the commander.

c. Execute *At Ease* and *Rest* with the guidon from the *Order Guidon* position.

Figure H-2. Parade Rest.

H-4. CARRY GUIDON

To assume the position of *Carry Guidon,* execute the following actions.

a. On the preparatory command for marching movements, execute *Carry Guidon* from *Order Guidon* by reaching across the body with the left hand (forearm horizontal) and grasping the staff (1, Figure H-3). Raise the staff vertically 6 inches allowing the staff to slide through the right hand. While the staff is sliding through the right hand, release the grasp and regrasp the staff with right thumb to the front, pointed downward and on line with the seam of the trousers. On the command of execution **MARCH**, return the left hand smartly to the left side and simultaneously step off (2, Figure H-3).

b. To resume *Order Guidon* from *Carry Guidon,* reach across the body with the left hand (forearm horizontal) and grasp the staff. Let the staff slide through the hands until the ferrule touches the marching surface. Regrasp the staff in the same manner as in *Order Guidon* and then return the left hand sharply to the side as in the position of attention.

c. When double-timing, carry the guidon diagonally across the body in the same manner as *Port Arms.*

Figure H-3. Carry position.

H-5. RAISED GUIDON

On the preparatory command ***Present*** of ***Present,*** **ARMS** and ***Eyes*** of ***Eyes,*** **RIGHT**, raise the guidon vertically, by grasping the staff with the right hand. Raise the guidon with the right hand while simultaneously moving the left hand (with palm to the rear) across the body to guide the guidon (1, Figure H-4). Keep raising the guidon until the right hand is on line with the right shoulder. Keep the right elbow into the side. Hold the guidon in this position until the command of execution is given (2, Figure H-4).

NOTE: When executing *Raised Guidon* from *Carry Guidon*, the bearer changes his grip to the *Order Guidon* position when he grasps the staff with his left hand. He then executes *Raised Guidon* as previously described. The bearer returns to *Carry Guidon* from *Raised Guidon* as described in paragraph H-4.

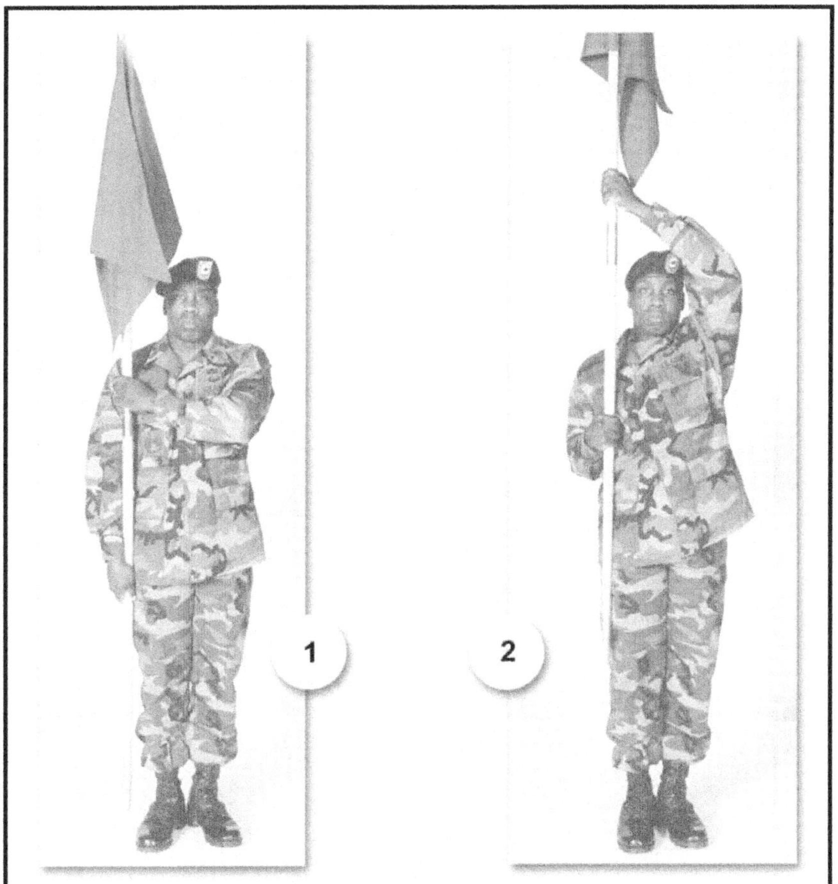

Figure H-4. Raised Guidon.

H-6. PRESENT GUIDON

To assume this position, execute the following actions.

a. On the preparatory commands for ***Present*** **ARMS** and ***Eyes*** **RIGHT**, execute *Raised Guidon.* On the command of execution, lower the guidon to the front using the right hand, keeping the left hand in the same position as *Raised Guidon* until the guidon is in the horizontal position resting under the arm pit. As soon as the staff is horizontal, return the left hand sharply to the left side (Figure H-5).

b. During a review, execute *Eyes Right* as the guidon is presented on the command of execution **RIGHT**. Move the head and eyes to the front as the guidon is raised on the command ***Ready*** of ***Ready***, **FRONT**.

c. On the command ***Order*** of ***Order***, **ARMS** or ***Ready*** of ***Ready***, **FRONT**, regrasp the staff with the left hand at its original position and execute *Raised Guidon.* On the command of execution **ARMS** or **FRONT**, lower the guidon back to the carry or order position.

Figure H-5. Present Guidon.

H-7. GUIDON BEARER'S SALUTE

When the guidon bearer (not in formation) has to salute from the order or carry position, the *Salute* is given with the left hand. To achieve this position, execute the following actions.

a. Move the left hand sharply to a position so that the first joint of the forefinger is touching the staff. The fingers and thumb are extended and joined, palm down, wrist straight, and forearm horizontal (Figure H-6).

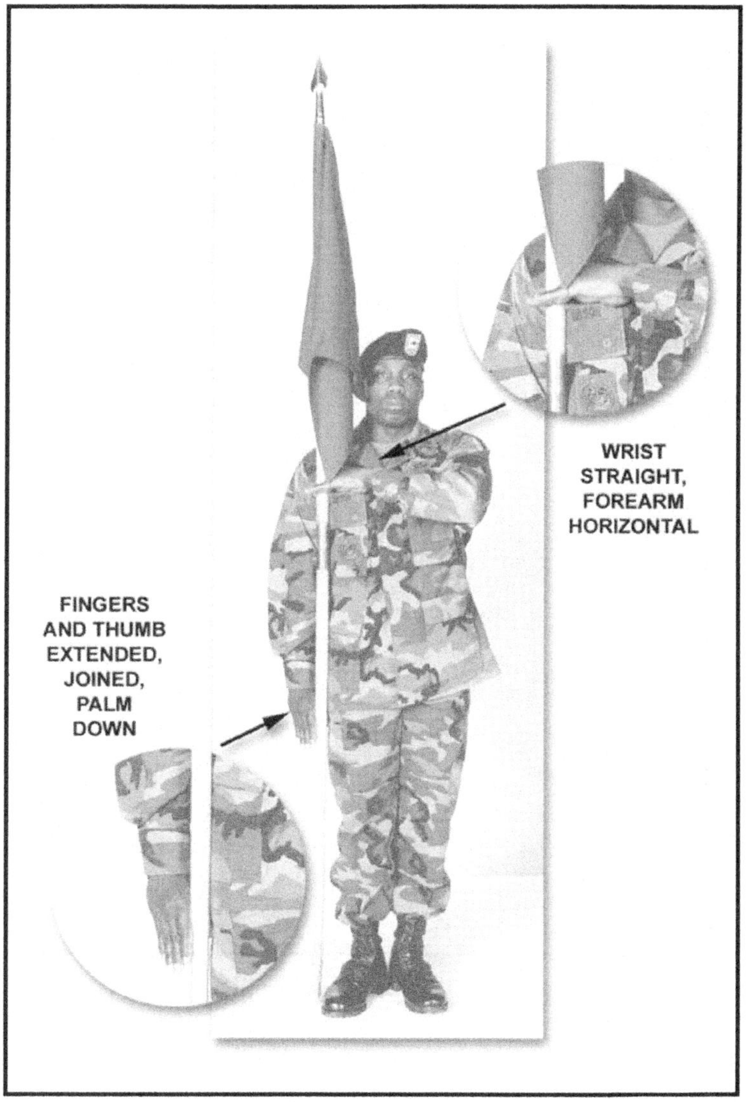

Figure H-6. Guidon Bearer's Salute.

b. After the *Salute* is acknowledged, return the left hand sharply to the side.

c. When at *Double Time* (not in formation), return to *Quick Time,* returning the guidon to the carry position and render the guidon *Salute* as previously described. Upon completing the *Salute* return to the carry position and resume *Double Time* (Figure H-7).

Figure H-7. Double Time.

APPENDIX I

CEREMONY CHECKLIST

This appendix contains a checklist for successfully conducting military ceremonies. It also shows the sequence of events for a change-of-command ceremony.

I-1. CHECKLIST

Figure I-1 shows a checklist for conducting military ceremonies.

____ **DETERMINE DATE/LOCATION**

____ **PREPARE/MAIL INVITATIONS**

____ **SUBMIT EXTERNAL SUPPORT REQUIREMENTS**

- ____ **Sound/podium**
- ____ **Seating**
- ____ **Programs**
- ____ **Coordinate with battalion staff**

____ **PREPARE INTERNAL MOI**

- ____ **Troop formation**
- ____ **Uniform/equipment**
- ____ **Mark field**
- ____ **Rehearsals**
- ____ **Ushers/seating plan**
- ____ **Guides/road guards**
- ____ **Flowers**
- ____ **Reception**
- ____ **Inclement weather plan**
- ____ **Guest list**
- ____ **Guidon and staff**

Figure I-1. Military ceremony checklist.

I-2. CHANGE OF COMMAND SEQUENCE OF EVENTS

Use the following sequence of events for a change-of-command ceremony.

- Prelude music.
- Formation of troops.
- Welcome (narrator).
- Introduction of official party (narrator).
- COT moves guidon forward.
- Official party moves to four steps from guidon.
- Official party/first sergeant takes position.
- Narrator reads assumption of command orders.
- Guidon is passed.
- COT returns guidon.
- Reviewing officer comments.
- Outgoing, then incoming, commander comments.
- Infantry/Army song.
- Narrator announces conclusion and administrative remarks.
- COT dismisses troops.

APPENDIX J
MOUNTED DRILL

This appendix contains the procedures for the orderly formation and movement of vehicles during drill and ceremonies for motorized and mechanized units. When necessary, the procedures may be modified to conform to local conditions.

J-1. GENERAL

Motorized and mechanized units, when dismounted, conduct drill and ceremonies in the same manner as prescribed for infantry units.

a. When mounted in vehicles, visible troops sit at *Attention.* Those armed with a rifle place the butt of the weapon alongside the right foot in the same manner as *Order Arms,* with the weapon vertical and parallel to the calf of the leg. Grasp the weapon with the right hand so that the right forearm is horizontal. Visible troops not armed with a rifle, fold their arms (right over the left) and raise the elbows so that the upper arms are horizontal.

b. When marching units and motorized and or mechanized units take part in the same ceremony, the units form dismounted (in a formation corresponding to the marching units) with their vehicles to the rear. On the directive **PASS IN REVIEW,** the commander faces about and directs at your vehicles, **FALL IN or MOUNT (LOAD) YOUR VEHICLES.** Subordinate commanders and leaders, in sequence from right to left, command **At your *vehicles,* FALL IN (MOUNT YOUR VEHICLES)** and all vehicle members double time to their vehicle. On the signal to mount, they mount their vehicle. Drivers observe their commander or leader for the signal to start engines.

J-2. FORMATIONS

The formations for motorized and mechanized units are basically the same as for infantry units. The interval between vehicles is about five steps. The distance is about 10 steps between vehicles one behind the other. Mass formations may be used for ceremonies or instruction by companies or larger units.

a. **Column**. In this formation, the vehicles are placed one behind the other, and the right edges of the vehicles are aligned.

b. **Line.** In this formation, vehicles are abreast, and the front edges of the vehicles are aligned.

c. **Mass.** A unit formed with two or more columns abreast constitutes a mass. Masses may be grouped as follows:

- LINE OF MASSES. Masses abreast with ten-step intervals (Figure J-1, page J-2).
- COLUMN OF MASSES. Masses placed one behind another with ten-step intervals between companies (Figure J-2, page J-3).
- LINE OF BATTALIONS IN LINE OF MASSES. Battalions in line of masses abreast with 20-step intervals between battalions.
- LINE OF BATTALIONS IN COLUMN OF MASSES. Battalions in column of masses abreast with 20-step intervals between battalions.

- COLUMN OF BATTALIONS IN LINE OF MASSES. Battalions in line of masses, placed one behind another, with 20-step intervals between battalions.
- COLUMN OF BATTALIONS IN COLUMNS OF MASSES. Battalions in column of masses, placed one behind another, with 20-step intervals between battalions.

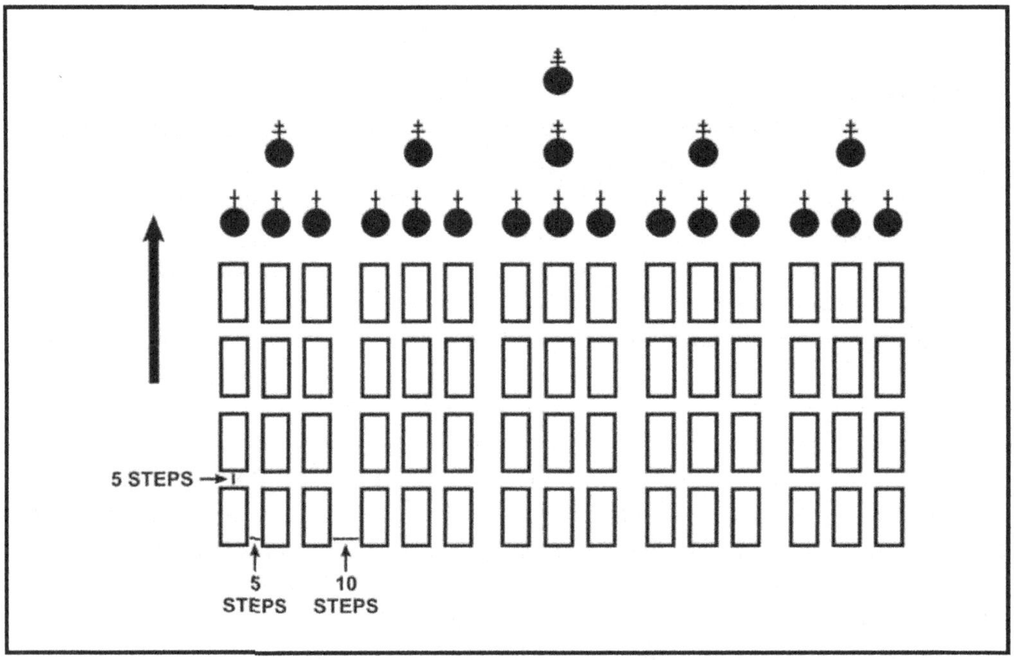

Figure J-1. Armor battalions, line of mass with vehicles in mass formation.

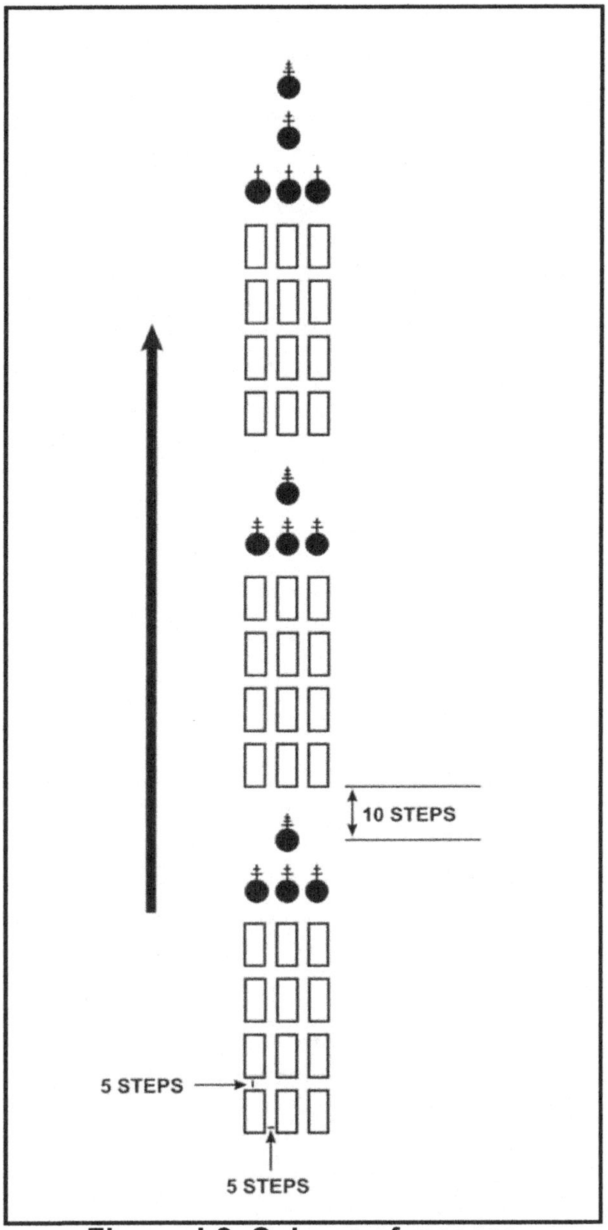

Figure J-2. Column of masses.

J-3. FORMATION AT VEHICLES

Personnel form at the vehicles before mounting using the following procedures.

a. To form at the vehicles from any formation, the command is ***At your vehicles,* FALL IN.** At the command **FALL IN,** the men move (double time) to their assigned vehicles and form as shown in Figure J-3 (page J-4). When there are more than five men, they form two or three ranks with normal distance. When there are more than 15 men, the number of men per rank may be increased. (Artillery sections form as prescribed in service-of-the-piece manuals.)

b. To load, the command is ***Load,*** **VEHICLES.** On the command of execution **VEHICLES,** each man double-times and forms at his vehicle (as previously described).

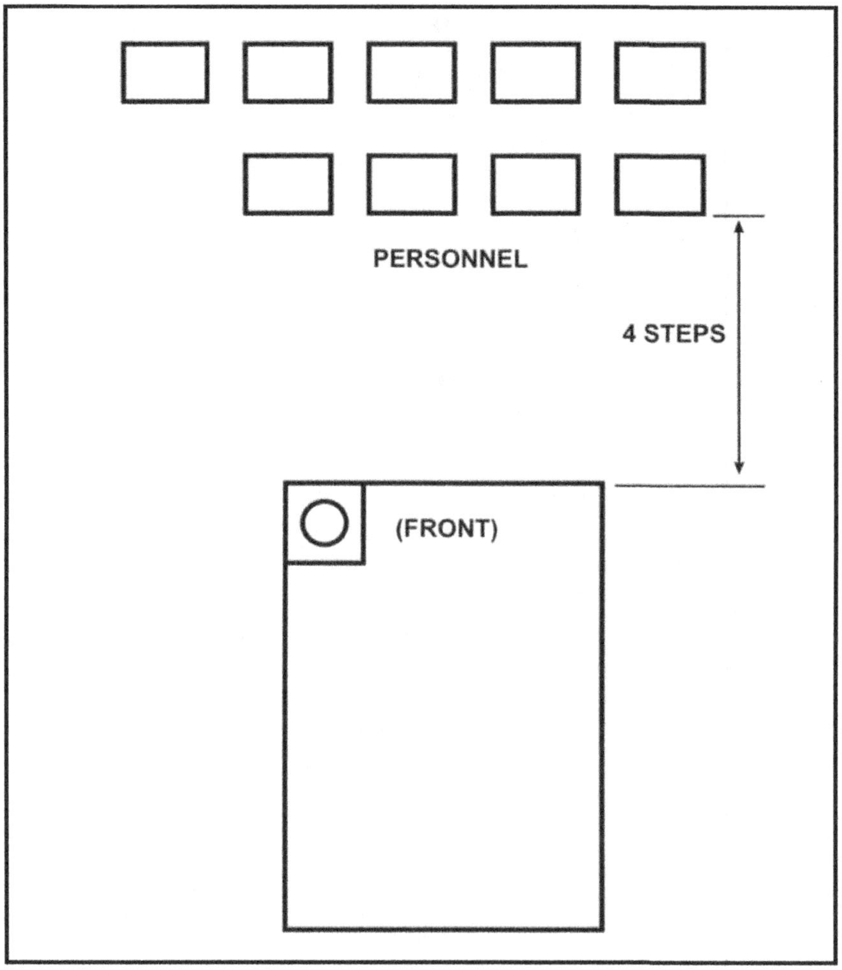

Figure J-3. Formation at the vehicle.

J-4. MOVEMENT OF VEHICLES

Execute the following actions to move vehicles.

a. To move forward, the commander gives the warning signal **FORWARD,** and his vehicle moves out. The other vehicles move forward at the designated distance. They are aligned on the base vehicle (right).

b. To halt, the commander signals **HALT.** The driver then slows down and halts the commander's vehicle. The other vehicles close to the prescribed distance and halt.

c. To change the direction of travel of mass formation, the commander gives the appropriate signal. On the commander's signal, the base vehicle turns 90 degrees and slows. Other vehicles turn about 45 degrees and continue traveling in an arc at normal speed until on line with the base vehicle. As the vehicles form on line, the base vehicle accelerates as necessary to maintain correct distance on the commander.

d. To change a line formation to a column formation, the commander signals for a column. The base vehicle, and those directly behind it, move forward. Other vehicles move at a 45-degree angle until in line with the lead vehicle and then execute a 45-degree turn.

e. To change a column formation to a line formation, the commander signals a line formation. The base (right flank vehicle) stops and all other vehicles turn at a 45-degree angle, turn again at a 45-degree angle when the prescribed interval is obtained, and then halt when on line with the base vehicle.

J-5. MOUNTED REVIEWS

Motorized and mechanized units conduct ceremonies as close as possible to the procedures prescribed for other units. They may participate (exactly as prescribed for marching units) as a dismounted unit, and on the directive **PASS IN REVIEW**, mount their vehicles and pass in review mounted. When desired, motorized and mechanized units may position their vehicles in a prescribed formation and remain mounted throughout the ceremony.

APPENDIX K
FLAGS AND COLORS

Normally, a flag detail consists of one noncommissioned officer in charge (NCOIC), two halyard pullers, and two to eight flag handlers. The purpose of the flag handlers is to ensure correct folding (unfolding) of the flag and to ensure that the flag does not touch the ground. As a guide, two flag handlers are needed when raising or lowering the storm (small) flag, six handlers for the post (medium) flag, and eight handlers for the garrison (large) flag.

The members of the flag detail are equipped according to local standing operating procedure and or letter of instructions.

The NCOIC inconspicuously gives the necessary commands or directives to ensure proper performance by the flag detail. On windy days, he may assist the flag handlers to secure or fold the flag. In most other situations, it is inappropriate for the NCOIC to touch the flag when it is being folded or unfolded.

NOTES: 1. For occasions when the flag of the United States is raised, lowered, or flown at half-staff, see AR 600-25 and paragraphs K-1 and K-2.

2. When the flag is to be flown at half-staff, it is first hoisted to the top of the flagpole and then lowered to the half-staff position. Before lowering the flag, it is again raised to the top of the flagpole and then lowered (AR 840-10).

K-1. RAISING THE FLAG (REVEILLE)

Execute the following actions when raising the flag.

a. The NCOIC forms the detail in a column of twos at *Double Interval* between files (Figure K-1, page K-2). He secures the flag from its storage area and positions himself between the files and on line with the last two men. He then marches the detail to the flagpole.

b. The detail is halted in column, facing the flagpole on the downwind side or as appropriate. They are halted so that the flagpole is centered between the halyard pullers. The NCOIC commands **POST.** On this command, the halyard pullers immediately move to the flagpole and ensure that the halyards are free of the pole. The flag handlers face to the center. The NCOIC then directs **UNFOLD THE FLAG.** On this directive, the two flag handlers nearest the NCOIC begin to (carefully) unfold the flag lengthwise, passing the freed end to the other handlers. When the two handlers nearest the flagpole have firmly secured the flag, the other handlers move away from the flagpole (as necessary) until the flag is fully extended. The flag is **not** unfolded widthwise. The flag handlers hold the flag waist high with their forearms horizontal to the ground.

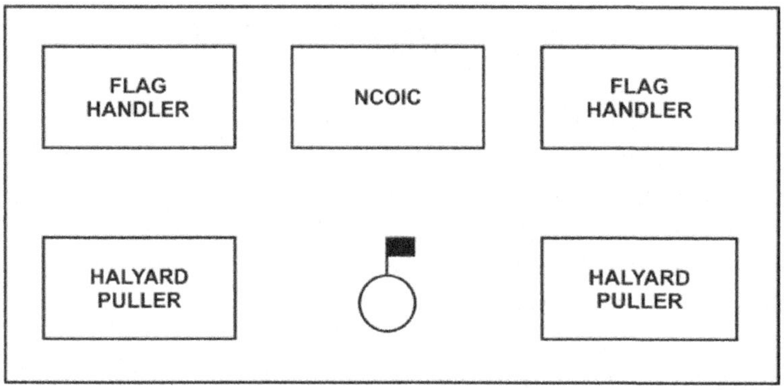

Figure K-1. Flag detail.

c. At the appropriate time, the NCOIC directs **ATTACH THE FLAG.** On this directive, all flag handlers take one side step toward the flagpole. The two handlers nearest the flagpole immediately attach the top of the flag to the halyard. The halyard pullers raise the flag until the bottom of the flag can be attached.

NOTE: When raising the larger flags, the NCOIC may command ***Ready,* *STEP*** while the halyard pullers raise the halyard until the lower portion of the flag is attached.

d. At the first note of the music, the reveille gun is fired and the halyard pullers rapidly raise the flag. The NCOIC salutes. As the flag is raised from the handlers' hands, they face the flagpole and salute. (Reveille is about 20 seconds in duration.) At the last note of the music, the NCOIC commands ***Order,*** **ARMS** for himself and the flag handlers while the pullers secure the halyards. The NCOIC then positions himself between the halyard pullers, executes an *About Face,* and commands ***Ready,*** **FACE.** The detail will face in the appropriate direction to depart the flag pole. He then marches the detail from the site.

K-2. LOWERING THE FLAG (RETREAT)

Execute the following actions when lowering the flag.

a. The detail is marched and positioned at the flagpole in the same manner as when raising the flag. On the command **POST,** the halyard pullers free the halyards, untangle them, ensure that they are free from the pole, and then temporarily resecure them; the flag handlers do not face to the center. The NCOIC then commands ***Parade,*** **REST.**

b. At the cannon shot or at the last note of "Retreat" (if a cannon is not fired), each halyard puller immediately frees the halyards. The NCOIC commands himself and the flag handlers to **ATTENTION** and ***Present,*** **ARMS.** At the first note of "To the Color," the flag is lowered slowly and with dignity.

c. The call, "To the Color," is approximately 40 seconds in duration. As the flag is lowered to within reach, the two flag handlers farthest away from the flag terminate their *Salute,* move forward rapidly, secure the flag, and move back from the flagpole between the columns.

(1) As the flag passes each handler, he terminates his *Salute* and assists in securing the flag. The flag is held palms down, fingers and thumb extended and joined, forearms horizontal. The NCOIC terminates his *Salute* at the last note of the music.

(2) Once the flag is detached, it is then folded. (See Figure K-2, page K-4, for correct folding techniques.) After securing the halyard, the handlers assist in the folding.

NOTE: When taking steps forward or backward to fold the flag, the members of the detail **always** step off with the left foot and **always** bring the trail foot alongside the lead foot as in the *Position of Attention.*

(a) To fold the flag the NCOIC commands ***Ready,*** **STEP.** All personnel take the appropriate number of steps backward to ensure that the flag is horizontal, wrinkle free, and centered on the flagpole.

(b) The NCOIC then commands ***Ready,*** **STEP.** The flag handlers take the appropriate number of steps toward each other, making the first fold lengthwise. The members on the NCOIC's right ensure that their edge of the flag is overlapping the left side by about 1 inch, which prevents any red from showing when the folding is complete. The NCOIC ensures that the flag is straight.

(c) The NCOIC then commands ***Ready,*** **TWO**. The members of the column on the NCOIC's left reach down and secure the fold with the fingers pointing upward and the palms facing away.

(d) The NCOIC then commands ***Ready,*** **THREE.** The members take the appropriate number of steps backward to ensure that the flag is horizontal, wrinkle free, and centered on the flagpole.

(e) ***Ready,*** **STEP,** ***Ready,*** **TWO,** and ***Ready,*** **THREE** are commanded and executed one more time to get the flag completely folded lengthwise.

(f) The flag handlers nearest the NCOIC then fold a 4-inch cuff and begin folding the flag, starting with the corner on the NCOIC's left. When folding the flag, the handlers are careful to keep the edges straight and to not bend the flag.

(g) After the flag has reached the position of honor (at the head of the right column), the NCOIC marches by the most direct route to a position directly in front of the flag handler and inspects the flag. The NCOIC then receives the flag and carries it held against his chest with his forearms with the point up.

(h) **After** the flag has been folded and received by the NCOIC, he positions himself between the halyard pullers and faces the storage site. The NCOIC then commands ***Ready,*** **FACE.** The members of the detail make the appropriate *Facing* movement toward the storage site. The detail is then marched to the storage site.

NOTE: Once the flag has been folded (cocked hat), it is treated as a cased Color and not saluted by persons meeting the flag detail. The flag will be treated with the utmost dignity and respect but not be rendered any sort of honors.

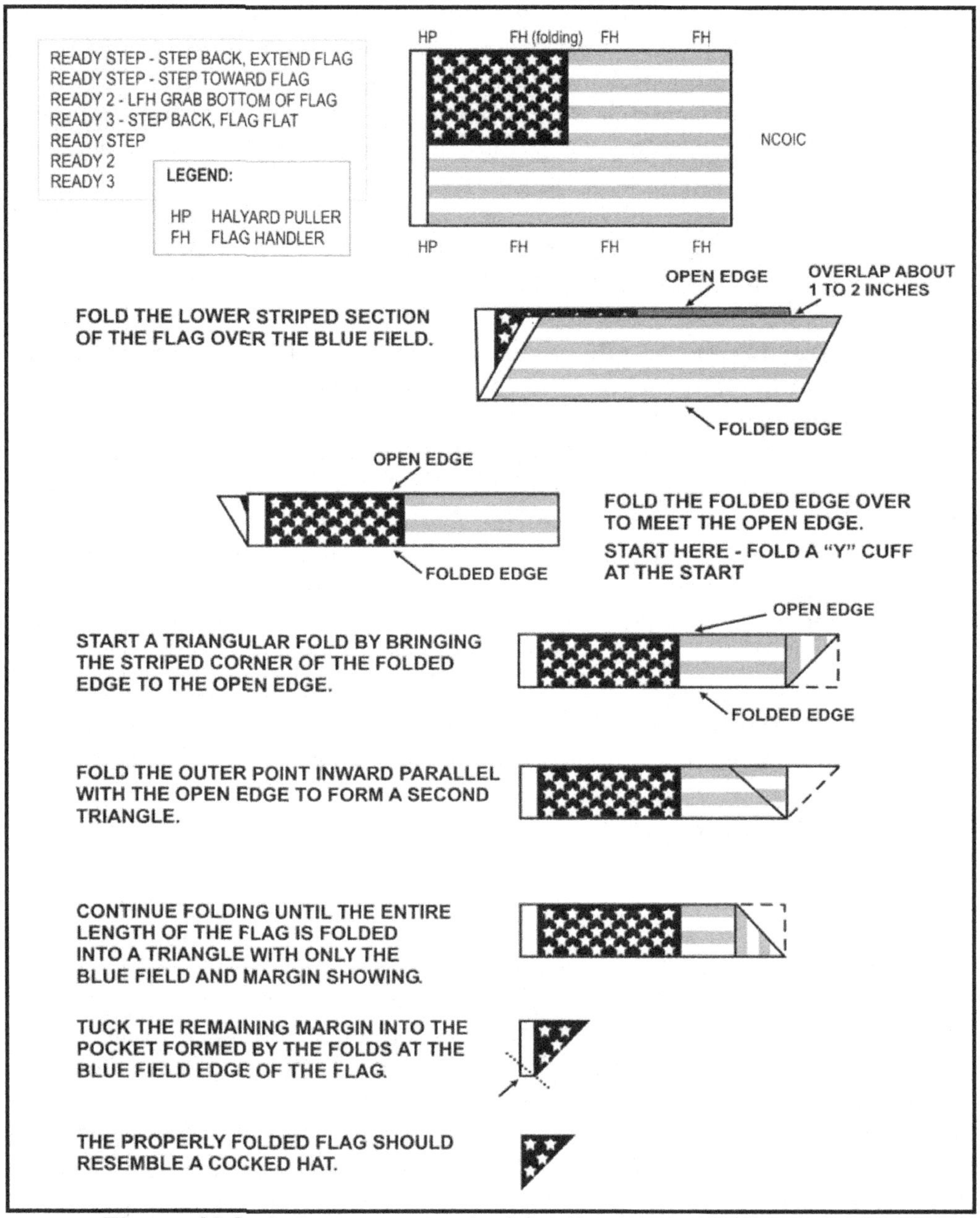

Figure K-2. Correct method of folding United States flag.

K-3. DISPLAY

The flag of the United States represents the living country and is considered a living thing.

a. Rules and customs for displaying the U.S. flag, adopted by an act of Congress in 1942 and amended in 1976, are found in Title 36 of the United States Code.

NOTE: No other flag or pennant should be placed above or, if on the same level, to the right of the flag of the United States of America except during church services conducted by naval chaplains at sea, when the church pennant may be flown above the flag, and except as required by multinational agreements to which the United States is a signatory.

b. The union, the field of blue with the grouping of stars representing the individual states, is the honor point and is the flag's right. When the flag is displayed, the union is always to the top and to the flag's own right (the left of the observer). When carried, the flag is always free, never stretched flat or carried horizontally. In the company of other flags, the U.S. flag is always on the marching right (the flag's own right), or, if there is a line of other flags, in front of the center of that line. When a number of flags of states, or localities, or pennants of societies are grouped and displayed from staffs, the U.S. flag is in the center and at the highest point of the group. If all flags are displayed from staffs of the same height, the flag of the United States is placed in the honor position, to the right of the other flags (to the left of the observer). When flags of two or more nations are displayed, they are flown from separate staffs of the same height, they are of approximately equal size, and the U.S. flag is to the right of the others.

NOTE: When the U.S. flag is flown in conjunction with other national flags, care must be taken to ensure that the foreign national flag is correct and properly displayed.

c. The flag of the United States is never used as drapery. It is displayed hung flat against a wall or flown free from a staff. It is not festooned over doorways or arches, tied in a bow, or fashioned into a rosette. It is not used to cover a speaker's desk or draped over the front of a platform. For those purposes, bunting may be used, giving the blue of the bunting the place of honor at the top of the arrangement or in the center of the rosette. Traditionally, there is one permissible departure from the rules for display of the flag of the United States: in a dire emergency, the flag may be flown upside down as a distress signal.

NOTE: For a detailed explanation of the regulations concerning the flag of the United States and other flags and Colors, see ARs 840-10 and 600-25.

K-4. CLASSIFICATION

A flag, as distinguished from a Color, is not saluted except during the ceremonies of raising and lowering the flag.

a. U.S. flags used by the Army are classified by size as follows:

(1) ***Garrison Flag***. The garrison flag is 20 feet hoist by 38 feet fly of nylon wool and is displayed on holidays and important occasions.

(2) ***Post Flag***. The post flag is 8 feet 11 3/8-inch hoist by 17 feet fly of nylon and is used for general display, and flown daily.

(3) ***Field Flag***. The field flag is 6 feet 8 inches hoist by 12 feet fly of nylon wool and is used for display with the positional field flag.

(4) ***Storm Flag***. The storm flag is 5 feet hoist by 9 feet 6 inches fly of nylon and is flown during inclement weather.

(5) ***Interment Flag***. The interment flag is 5 feet hoist by 9 feet 6 inches fly of cotton bunting.

b. A Color is the U.S. flag trimmed on three sides with golden yellow fringe 2 1/2 inches wide. The Color is primarily for indoor display. U.S. Colors used by the Army are classified by size as follows:

- 4 feet, 4 inches hoist by 5 feet, 6 inches fly displayed with the U.S. Army flag, positional Colors, the Corps of Cadets Colors, 1st Battalion 3rd Infantry Colors, and the chapel flag.
- 3 feet hoist by 4 feet fly displayed with the Army field flag, distinguishing flags, organizational Colors, institutional flags, and the chapel flag.

c. Other types of military flags include:

(1) ***Colors***. These are flags of organizations and certain civilian and military officials.

(2) ***Standards***. Flags of mounted units were formerly called standards.

(3) ***Distinguished Flags.*** These flags identify headquarters, offices, general officers, and organizations, none of which is authorized organizational or individual Colors.

(4) ***Ensign.*** An ensign is a rectangular flag flown from aircraft, ships, and boats.

(5) ***Guidon***. A guidon is a swallow-tailed flag carried by companies, batteries, troops, and certain detachments.

(6) ***Pennant***. A pennant is a triangular flag used for various utility purposes.

d. Dismounted organizations traditionally carried the Color while mounted organizations (cavalry, mechanized, and motorized) traditionally carried the standard. Colors were larger than standards. Under present regulations, both are called Colors.

e. The organizational Color is dipped in *Salute* during the playing of the United States National Anthem, "To the Color," "Taps" (funerals only), and during the ceremony at reveille. It is dipped in *Salute* when rendering honors to the organization's commander or to a person of higher grade, but at no other time. The U.S. flag is never dipped in *Salute.*

f. Until 1813, the Colors of an infantry regiment were carried by ensigns, who were the lowest ranking officers of the regiment. In that year, the regulations were changed and the work was entrusted to Color sergeants. Noncommissioned officers carry them today.

K-5. ARMY FLAGS

The United States Army flag is 4 feet, 4 inches hoist by 5 feet, 6 inches fly and is trimmed on three sides with yellow fringe 2 1/2 inches in width. On Flag Day, 14 June 1956, the 181st anniversary of the U.S. Army, the newly adopted United States Army flag was publicly unfurled at Independence Hall, Philadelphia, Pennsylvania, by the Honorable Wilbur M. Brucker, Secretary of the Army. The flag, of white silk, bears an embroidered replica of the official seal of the Department of the Army in ultramarine blue (without the roman numerals). A scarlet scroll inscribed "United States Army" in white is centered between the device and the ultramarine blue numerals "1775" denoting the year the Army was founded, by action of the Continental Congress, 14 June 1775. The original War Office seal, constituting the central design of the flag, was authorized by the Continental Congress on 8 May 1779. The U.S. Army flag bears all the streamers representing the Army's campaigns since its inception. (When not being carried, the Yorktown streamer should always be prominently displayed.)

The Army field flag was authorized in 1962 and is the same as the United States Army flag except that it is 3 feet hoist by 4 feet fly, the background is ultramarine blue, the seal is white, the scroll is white, "United States Army" is scarlet, the numerals "1775" are white, and streamers are not authorized. The flag is issued to those organizations and headquarters not authorized the United States Army flag. (Units authorized issue are specified in AR 840-10.)

K-6. COURTESIES BY INDIVIDUALS

Table K-1 (pages K-8 through K-11) lists courtesies by individuals in various dress and circumstances.

COURTESIES BY INDIVIDUALS
(ADAPTED FROM AR 600-25, APPENDIX A)

1. MILITARY PERSONNEL IN UNIFORM (WITH OR WITHOUT HEADDRESS) AND IN FORMATION.

REVEILLE	RETREAT WHEN *PLAYED AS PRELUDE* TO "TO THE COLOR"	"TO THE COLOR" OR *NATIONAL ANTHEM*	WHEN UNCASED *COLORS PASS BY* OR WHEN PASSING UNCASED COLORS	CANNON *SALUTES RENDERED* AS HONOR TO A PERSON (NOTE 1)	MILITARY *FUNERALS*
Execute *Present Arms* and *Order Arms* at the command of OIC or NCOIC.	Execute *Parade Rest* at the command of OIC or NCOIC. Remain at *Parade Rest* until given *Attention* by OIC or NCOIC.	Same as for *Reveille.*	COLORS ARE PASSING: Execute *Present Arms* at the command of OIC or NCOIC. Command is given when Colors are within six paces of the flank of the unit, and the salute is held until the Colors are six paces past the flank of the unit, at which time the command *Order Arms* is given. PASSING THE COLORS: Execute orders at the command of OIC or NCOIC.	Execute *Present Arms* at the command of OIC or NCOIC. Hold *Salute* until command *Order Arms* is given by OIC or NCOIC.	Execute *Present Arms* at the command of OIC or NCOIC.

Notes:
1. When the cannon salute to the Union or nation is fired, no individual action is required.
2. Military photographers covering ceremonies will render appropriate courtesy during the playing of the National Anthem.

Table K-1. Courtesies by individuals.

COURTESIES BY INDIVIDUALS					
2. MILITARY PERSONNEL IN UNIFORM (WITH OR WITHOUT HEADDRESS) NOT IN FORMATION.					
REVEILLE	**RETREAT WHEN PLAYED AS PRELUDE TO "TO THE COLOR"**	**"TO THE COLOR" OR NATIONAL ANTHEM**	**WHEN UNCASED COLORS PASS BY OR WHEN PASSING UNCASED COLORS**	**CANNON SALUTES RENDERED AS HONOR TO A PERSON (NOTE 1)**	**MILITARY FUNERALS**
At the first note of music, face the flag (or music if flag is not in view) and render *Hand Salute*. End *Salute* on last note of music.	At the first note of music, face the flag (or music if flag is not in view) and stand at *Attention*. Hold that position until the last note of "Retreat" has been played.	(Note 2) At the first note of music: if outdoors, render *Hand Salute*; if indoors, stand at *Attention*. Hold that position until last note of music has been played.	COLORS ARE PASSING: When colors are within six paces: if outdoors, render *Hand Salute*; if indoors, stand at *Attention*. Hold that position until Colors have passed six paces. PASSING THE COLORS: Outdoors: when within six paces of the colors, turn head in direction of the Colors and render *Hand Salute*. Hold *Salute* until six paces past the Colors.	At the first note of music or first round of *Salute*, face the ceremonial party and render *Hand Salute*. End *Salute* on last note of music or when last round of *Salute* has been fired.	Each time casket is moved: if outdoors, render *Hand Salute*; if indoors, stand at *Attention*.

Table K-1. Courtesies by individuals (continued).

COURTESIES BY INDIVIDUALS

3. ALL MEN (CIVILIAN AND MILITARY) WEARING CIVILIAN CLOTHERS (INCLUDING SPORTS UNIFORM) WITH HEADDRESS.

REVEILLE	RETREAT WHEN PLAYED AS PRELUDE TO "TO THE COLOR"	"TO THE COLOR" OR NATIONAL ANTHEM	WHEN UNCASED COLORS PASS BY OR WHEN PASSING UNCASED COLORS	CANNON SALUTES RENDERED AS HONOR TO A PERSON (NOTE 1)	MILITARY FUNERALS
At the first note of music, face the flag (or music if flag is not in view), stand at *Attention*, and remove headdress with right hand and hold over left shoulder with right hand over the heart. Hold that position until last note of music has been played.	At the first note of music, face the flag (or music if flag is not in view), remove headdress with right hand, and stand at *Attention*. Remain at *Attention* until last note of "Retreat" has been played.	(Note 2) At the first note of music: if outdoors, hold headdress over the left shoulder with the right hand over the heart; if indoors, stand at *Attention*. Hold that position until last note of music has been played.	COLORS ARE PASSING: When colors are within six paces: if outdoors, stand at *Attention*, remove headdress with right hand, and hold over the left shoulder with right hand over heart; if indoors, stand at *Attention*. Hold that position until Colors have passed six paces. PASSING THE COLORS: Outdoors, when within six paces of Colors, turn head in direction of Colors, remove headdress with right hand, and hold over left shoulder with right hand over the heart. Hold that position until six paces past the colors.	At the first note of music or first round of *Salute*, face the ceremonial party, remove headdress, and stand at *Attention*. Hold that position until last note of music or last round of *Salute* has been fired.	Each time casket is moved: if outdoors, stand at *Attention*, remove headdress with right hand, and hold over the left shoulder with right hand over heart; if indoors, stand at *Attention*.

Table K-1. Courtesies by individuals (continued).

COURTESIES BY INDIVIDUALS					
4. MILITARY PERSONNEL AND CIVILIANS IN CIVILIAN DRESS WITHOUT HEADDRESS. FEMALE PERSONNEL (CIVILIAN AND MILITARY NOT IN UNIFORM) WITH OR WITHOUT HEADDRESS. PERSONNEL ENGAGED IN SPORTS AND ATTIRED IN SPORT UNIFORM WITHOUT HEADDRESS.					
REVEILLE	**RETREAT WHEN PLAYED AS PRELUDE TO "TO THE COLOR"**	**"TO THE COLOR" OR NATIONAL ANTHEM**	**WHEN UNCASED COLORS PASS BY OR WHEN PASSING UNCASED COLORS**	**CANNON SALUTES RENDERED AS HONOR TO A PERSON (NOTE 1)**	**MILITARY FUNERALS**
At the first note of music, face the flag (or music if flag is not in view) and stand at *Attention* with right hand over heart. Hold that position until last note of music has been played.	At the first note of music, face the flag (or music if flag is not in view) and stand at *Attention*. Remain at *Attention* until last note of "Retreat" has been played.	(Note 2) At the first note of music: if outdoors, stand at *Attention* with right hand over heart; if indoors, stand at *Attention*. Hold that position until last note of music has been played.	COLORS ARE PASSING: When colors are within six paces: if outdoors, stand at *Attention*, with right hand over heart; if indoors, stand at *Attention*. Hold that position until Colors have passed six paces. PASSING THE COLORS: No action is required.	At the first note of music or first round of *Salute*, face the ceremonial party and stand at *Attention*. Hold that position until last note of music or last round of *Salute* has been fired.	Each time casket is moved: if outdoors, stand at *Attention* with right hand over heart; if outdoors, stand at *Attention*.

Table K-1. Courtesies by individuals (continued).

This page intentionally left blank.

GLOSSARY

Acronyms and Abbreviations

AR	Army Regulation
ARNG	Army National Guard
AWOL	absent without leave
*CAC	Casualty Assistance Center
CAO	casualty assistance officer
CD	compact disk
COT	commander of troops
CQ	charge of quarters
DOD	Department of Defense
FM	field manual
MOI	memorandum of instruction
NCO	noncommissioned officer
NCOIC	noncommissioned officer in charge
NOK	next of kin (graphics only)
OIC	officer in charge
OSD	Office of the Secretary of Defense; over, short, and damaged report
POC	point of contact
Pvt	private
RC	Reserve Component
RSC	Regional Support Command
SOP	standing operating procedures
STARC	state area command
U.S.	United States
USAR	U.S. Army Reserve

Definitions

alignment	The arrangement of several elements on the same line
base	The element on which a movement is planned or regulated.

cadence The uniform rhythm in which a movement is executed, or the number of steps or counts per minute at which a movement is executed. Drill movements are normally executed at the cadence of quick time or double time. Quick time is the cadence of 120 counts or steps per minute; double time is the cadence of 180 counts or steps per minute.

ceremonies Formations and movements in which a number of troops execute movements in unison and with precision just as in drill; however, their primary value is to render honors, preserve tradition, and stimulate esprit de corps.

commander Person in charge.

cordon A line of soldiers to honor a dignitary upon entering or exiting from a given place or vehicle.

cover Aligning oneself directly behind the man to one's immediate front while maintaining correct distance.

depth The space from front to rear of a formation, including the front and rear element.

directive An oral order given by a commander to direct or cause a subordinate leader or lead element to take action.

distance The space between elements when the elements are one behind the other. Between units, it varies with the size of the formation; between individuals, it is an arm's length to the front plus 6 inches, or about 36 inches, measured from the chest of one man to the back of the man immediately to his front.

drill Certain movements by which a unit (or individuals) is moved in a uniform manner from one formation to another, or from one place to another. Movements are executed in unison and with precision.

element An individual, squad, section, platoon, company, or larger unit forming as part of the next higher unit.

file A column that has a front of one element.

flank The right or left side of any formation as observed by an element within that formation.

formation The arrangement of elements of a unit in a prescribed manner:

line — A formation in which the elements are side by side or abreast of each other. In a platoon line, the members of each squad are abreast of each other with the squads one behind the other.

column — A formation in which the elements are one behind the other. In a platoon column, the members of each squad are one behind the other, with the squads abreast of each other. To change a line formation to a column formation, the command is *Right,* FACE. To change a column formation to a line formation, the command is *Left,* FACE.

front — The space from side to side of a formation, including the right and left elements.

guide — The person responsible for maintaining the prescribed direction and rate of march.

head — The leading element of a column.

interval

- close — The lateral space between soldiers, measured from right to left by the soldier on the right placing the heel of his left hand on his hip, even with the top of the belt line, fingers and thumb joined and extended downward, with his elbow in line with the body and touching the arm of the soldier to his left.
- double — The lateral space between soldiers, measured from right to left by raising both arms shoulder high with the fingers extended and joined (palms down) so that fingertips are touching the fingertips of the soldiers to the right and to the left.
- normal — The lateral space between soldiers, measured from right to left by the soldier on the right holding his left arm shoulder high, fingers and thumb extended and joined, with the tip of his middle finger touching the right shoulder of the soldier to his left.

PICAA — five-step process used in all marching movements: P-preparatory command, I-intermediate step, C-command of execution, A-action step, and A-additional step

post — The correct place for an officer or noncommissioned officer to stand in a prescribed formation.

rank — A line that is one element in depth.

re-form — A command to restore the previous element or formation (used only during drill instructions).

step — The prescribed distance measured from one heel to the other heel of a marching soldier.

This page intentionally left blank.

REFERENCES

*READINGS RECOMMENDED

These readings contain relevant supplemental information.

*AR 220-20 Army Status of Resources and Training Systems (ASORTS)—Basic Identity Data Elements (BIDE). 19 March 2004.

*AR 310-25 Dictionary of United States Army Terms. 15 October 1983. (obsolete)

*AR 600-25 Salutes, Honors, and Visits of Courtesy. 24 October 2004.

*AR 670-1 Wear and Appearance of Army Uniforms and Insignia. 02 March 2005.

AR 840-10 Flags, Guidons, Streamers, Tabards, and Automobile and Aircraft Plates. 01 November 1998.

*DA Pamphlet 638-2 Care and Disposition of Remains and Disposition of Personal Effects. 22 December 2000.

*FM 3-23.35 Combat Training with Pistols, M9 and M11. 25 June 2003; Change 1, 5 September 2005.

*INTERNET WEBSITES

U. S. Army Publishing Directorate, http://www.usapa.army.mil

Army Doctrine and Training Digital Library, http://www.adtdl.army.mil

INDEX

FM 3-21.5 (FM 22-5)
7 JULY 2003

By Order of the Secretary of the Army:

JOHN M. KEANE
General, United States Army
Acting Chief of Staff

Official:

JOEL B. HUDSON
Administrative Assistant to the
Secretary of the Army
0317804

DISTRIBUTION:

Active Army, Army National Guard, and U. S. Army Reserve: To be distributed in accordance with the initial distribution number 110178, requirements for FM 3-21.5.

PIN: 080918-001

Made in the USA
Coppell, TX
21 July 2026

85268178R00164